Noncoding RNA - The Dark Matter of the Genome

Edited by Preeti Dabas

Published in London, United Kingdom

Noncoding RNA - The Dark Matter of the Genome
http://dx.doi.org/10.5772/intechopen.111315
Edited by Preeti Dabas

Contributors
Aleksandr Abramov, Ali Afzal, Alisa Petkevich, Assaf C. Bester, Cheng Chen, Christos Kotakis, Huma Rasheed, Jajati Keshari Nayak, Le Qu, Maria Radanova, Meijuan Wu, Muddasir Hassan Abbasi, Muhammad Babar Khawar, Nadeem Sheikh, Rashmi Chauhan, Sharat Prabhakaran, Sundip Kumar, Vadim Pospelov, Yulin Zhou

First published in London, United Kingdom, 2024 by IntechOpen
IntechOpen is the global imprint of INTECHOPEN LIMITED, registered in England and Wales, registration number: 11086078, 167-169 Great Portland Street, London, W1W 5PF, United Kingdom

British Library Cataloguing-in-Publication Data
A catalogue record for this book is available from the British Library

Additional hard and PDF copies can be obtained from orders@intechopen.com

Noncoding RNA - The Dark Matter of the Genome
Edited by Preeti Dabas
p. cm.

This title is part of the Genetics Book Series, Volume 2
Topic: Genomics
Series Editor: Kenji Ikehara
Topic Editor: Irina Vlasova-St. Louis

Print ISBN 978-1-80356-710-5
Online ISBN 978-1-80356-711-2
eBook (PDF) ISBN 978-1-80356-712-9
ISSN 3049-7094

For EU product safety concerns:
IN TECH d.o.o., Prolaz Marije Krucifikse Kozulić 3, 51000 Rijeka, Croatia,
info@intechopen.com or visit our website at intechopen.com.

IntechOpen Book Series
Genetics
Volume 2

Aims and Scope of the Series

"Genetics," which has been proud of its tradition since Mendel presented his research results in 1865, initially progressed quite slowly due to simple observational approaches of individuals and groups. However, the discovery of double-stranded DNA by Watson and Crick about 70 years ago triggered rapid progress in life sciences, including genetics, which was primarily conducted using Escherichia coli and bacteriophages infecting E. coli. Subsequently, genetics has achieved remarkable developments, such as understanding genetic disorders, including cancers, through research on the biogenesis and differentiation of plants and animals.
The two topics of this book series - Human Genetics, and Genomics - will address important areas of advancement in genetics.
Human Genetics: After fundamental genetics, initially studied with the main goal of revealing the functions of individual genes and proteins, genetics expanded from understanding the genetic system itself to understanding many infectious diseases caused by bacteria and viruses. Consequently, human beings are now overcoming infectious diseases by developing medicinal chemicals, including antibiotics and vaccines. However, genetic disorders remain challenging to cure up to now. Nevertheless, even the cure for them, including various cancers, is coming closer to reality due to the rapid progress of human genetics. In this way, the welfare of human life continues to improve, and even longevity, which was once a dream, has been achieved to some extent in recent years.
Genomics: On the other hand, the understanding of the comprehensive interrelationship of whole genes or whole proteins functioning in one organism has become possible now, as research has entered the era of genomics, owing to the rapid progress of base sequence analysis and bioinformatics. The development of genomics has further made it possible to understand the evolutionary processes of organisms through comparative studies among the genomes of many organisms.
This book series will discuss the findings obtained during the advancement of human genetics and genomics. It is also expected that this series will trigger the formation of a better world composed of human beings and all other organisms on Earth through discussions of research results obtained under the development of general genetics.

Meet the Series Editor

Kenji Ikehara graduated from the Department of Industrial Chemistry, Faculty of Engineering, Kyoto University in 1968. He received his B. Eng. (1968) and subsequently earned M. Eng. (1970) and D. Eng. (1976) degrees from Kyoto University. He began his career as a research associate in the Faculty of Science at the University of Tokyo before moving on to become an associate professor in the Faculty of Science at Nara Women's University. He was later promoted to professor and subsequently served as the dean of the Faculty of Science at Nara Women's University. Additionally, he held the position of director at the Nara Study Center of the Open University of Japan. For approximately 15 years, he focused his research on sporulation initiation of Bacillus subtilis. Later, he shifted his focus to the origins and evolutionary processes of microbial genes, the genetic code, proteins, and life. He has proposed several hypotheses, including the GC-NSF(a) hypothesis on the origin of genes, the GNC-SNS hypothesis on the genetic code, the protein 0th-order structure hypothesis on the origin of proteins, and the [GADV]-protein world hypothesis (GADV hypothesis) on the origin of life. Furthermore, he served as the local chair of the International Conference, Origin 2014, held in Nara in 2014.

Meet the Volume Editor

Dr. Preeti Dabas, a motivated scientific professional, grew up in India and earned her Ph.D. from Guru Gobind Singh Indraprastha University in Delhi, India, where she focused on yeast genetics, molecular biology, biochemistry, and biophysical techniques. She conducted her postdoctoral research at St. Jude Children's Research Hospital in Memphis, USA, specializing in gene regulation, signal transduction, and hematopoiesis crucial for understanding development and disease, particularly cancer and blood disorders. Dr. Dabas also gained valuable experience as a junior research fellow at the Indian Institute of Advanced Research in India and as a researcher at the Institute of Biomedical Technology in Finland.

Contents

Preface **XIII**

Section 1
Regulatory Roles of Non-Coding RNA **1**

Chapter 1 **3**
Deciphering the Hidden Language of Long Non-Coding RNAs: Recent Findings and Challenges
by Assaf C. Bester

Chapter 2 **27**
Transcribed-Ultra Conserved Regions (T-UCRs) a New Light on a Dark Matter
by Maria Radanova

Chapter 3 **49**
LncRNAs in Cancer Development
by Alisa Petkevich, Aleksandr Abramov and Vadim Pospelov

Chapter 4 **69**
Circulating lncRNAs in Cancer
by Le Qu, Cheng Chen, Meijuan Wu and Yulin Zhou

Section 2
Noncoding RNA in Development **89**

Chapter 5 **91**
Shedding Light on the Dark Matter, Noncoding RNAs in Human Reproduction
by Rana Alhamdan

Chapter 6 **113**
Functional Role of MicroRNAs in Embryogenesis
by Muhammad Babar Khawar, Huma Rasheed, Ali Afzal, Muddasir Hassan Abbasi and Nadeem Sheikh

Chapter 7 **147**
Photosynthesis in Plants Undergoing Silencing
by Christos Kotakis

Chapter 8 **165**
Whispers of Nourishment: Unveiling the Role of Non-Coding RNA in Plant Nutrient Availability
by Jajati Keshari Nayak, Rashmi Chauhan, Sundip Kumar and Sharat Prabhakaran

Preface

The genome, that intricate tapestry of genetic information, has long been the focus of intense scientific scrutiny. Yet, within its seemingly well-defined boundaries of coding genes lies a vast and mysterious realm—the dark matter of the genome. In recent years, the exploration of noncoding RNA (ncRNA) has revealed an intricate and previously underestimated layer of genetic regulation. These ncRNAs, though not coding for proteins, play crucial roles in a myriad of cellular processes and developmental stages. Our understanding of these molecules is still evolving, yet the insights gained thus far have profound implications for both basic biology and applied biomedical research. This book, *Dark Matter of the Genome*, aims to consolidate and expand upon current knowledge, offering a comprehensive examination of the regulatory roles and developmental functions of noncoding RNAs.

Long noncoding RNAs (lncRNAs) are key players in various cellular processes, yet the mechanisms behind their emergence and functions are not fully understood. Section 1, "Regulatory Roles of Non-Coding RNA", embarks on a journey to uncover these hidden mechanisms and regulatory functions of ncRNAs. The first chapter, "Deciphering the Hidden Language of Long Non-Coding RNAs: Recent Findings and Challenges", delves into the intricate world of ncRNA-mediated gene regulation. From microRNAs (miRNAs) to lncRNAs, these molecules communicate in a language of sequences, secondary structures, and binding motifs, exerting precise control over gene expression programs. Moving forward, the second chapter, "Transcribed-Ultra Conserved Regions (T-UCRs) a New Light on a Dark Matter" sheds light on a unique class of ncRNAs derived from highly conserved DNA regions and their regulatory roles. The third chapter. "LncRNA in Cancer", explores the role of lncRNAs in cancer pathogenesis, progression, and therapeutic resistance. From oncogenic lncRNAs that drive tumorigenesis to tumor-suppressive lncRNAs that modulate cellular senescence and apoptosis, the landscape of ncRNAs in cancer biology is as diverse as it is dynamic. On the other hand, the next chapter, "Circulating lncRNAs in Cancer" discusses another class of noncoding RNA termed as circulating lncRNAs, once dismissed as cellular debris. Circulating lncRNAs have recently emerged as promising diagnostic and prognostic markers for various cancers.

Section 2, "Noncoding RNA in Development", explains the pivotal roles of ncRNAs in shaping biological development and progression. The fifth chapter, "Shedding Light on the Dark Matter, Noncoding RNAs in Human Reproduction" discusses the involvement of specific ncRNAs in spermatogenesis, folliculogenesis, and oocyte maturation, along with their links to reproductive disorders. Furthermore, the sixth chapter, "Functional Role of MicroRNAs in Embryogenesis" provides an overview of biogenesis and its role during embryonic development, where precise temporal and spatial regulation is essential for orchestrating cellular differentiation, tissue morphogenesis, and organogenesis. Shifting focus to plant biology, the next chapter, "Photosynthesis in Plants Undergoes Silencing", investigates the fascinating interplay between ncRNAs and photosynthetic processes in plants. Moreover, the eighth chapter, "Whispers of Nourishment: Unveiling the Role of Non-Coding RNA in Plant Nutrient Availability" examines the roles of various ncRNAs in nutrient sensing, uptake, and utilization in plants. From microRNA-guided nutrient transport to lncRNA-mediated nutrient signaling pathways, plants employ a sophisticated ncRNA toolkit to adapt to varying nutrient availabilities and optimize growth and development.

As we navigate through the chapters of this book, we discover the fascinating diversity and complexity of ncRNA biology, from regulatory intricacies to developmental milestones and disease landscapes. Each chapter highlights the significant influence of ncRNAs, illuminating their roles as molecular architects of life and disease. *Noncoding RNA – The Dark Matter of the Genome* is a scientific journey that serves as a foundation for future research, fostering a deeper understanding of ncRNAs and their potential to revolutionize diagnostics and therapeutics.

Preeti Dabas
St. Jude Children's Research Hospital,
Memphis, USA

Section 1

Regulatory Roles of Non-Coding RNA

Chapter 1

Deciphering the Hidden Language of Long Non-Coding RNAs: Recent Findings and Challenges

Assaf C. Bester

Abstract

Long non-coding RNAs (lncRNAs) are crucial non-coding RNA genes involved in diverse cellular processes. However, the mechanisms underlying their emergence and functions remain incompletely understood. A major challenge in the field is to understand how lncRNA sequences affect their function. In recent years, comprehensive genetic and genomic studies have started to unfold the function of lncRNAs through their interactions, cellular organization, and structure. This comprehensive review delves into the intricate interplay between lncRNA sequences and their functional implications. Unlike other RNA types, lncRNAs exhibit a complex syntax, employing diverse functional elements such as protein recognition and miRNA binding sites, repeat elements, secondary structures, and non-canonical interactions with RNA and DNA binding proteins. By unraveling the hidden language that governs the function and classification of lncRNAs, we aim to shed light on the underlying principles shaping their diverse functions. Through a detailed examination of the intricate relationship between lncRNA sequences and their biological effects, this review offers insights into the sequences underlying lncRNA functionality. Understanding the unique sequence characteristics and functional elements employed by lncRNAs has the potential to advance our knowledge of gene regulation and cellular processes, providing a foundation for the development of novel therapeutic strategies and targeted interventions.

Keywords: lncRNA, ncRNA, RNA regulation, RNA motifs, RNA binding proteins

1. Introduction

Long non-coding RNAs (lncRNAs) represent a substantial group of non-protein-coding genes (PCGs) in the human genome, with at least 20,000 unique genes [1]. These transcripts typically exceed 200 nucleotides (nt) in length, possess minimal or no open-reading frame (ORF) potential, and do not belong to any other recognized RNA groups.

lncRNAs can be classified based on their genetic context in relation to PCGs (intergenic, divergent, intronic, and antisense), or their mechanism of action, functioning either locally (cis) or remotely from the transcription site (trans). However, unlike other types of RNAs (e.g., mRNA, miRNA, rRNA, tRNA, and snoRNA) where the function is clearly embodied in the sequence, the sequence-to-function relationship of lncRNAs is not as clear. This complexity forms the core focus of this manuscript.

Some lncRNAs function in a sequence-dependent manner, interacting specifically with other macromolecules, including RNA, DNA, and proteins (e.g., NORAD, NEAT1, and MALAT1). Others may rely solely on their processing, such as the initiation of their transcription (e.g., PVT1 [2]) or their splicing (e.g., Blustr [3]). In these instances, the specific sequence of the mature transcript is not essential. These lncRNAs often regulate PCGs in close proximity (cis-regulation) or overlap with DNA regulatory elements, such as enhancers (**Figure 1**) [4, 5].

While lncRNAs are characterized by having minimal or no ORF potential, some have been found to translate into functional peptides [6, 7]. However, lncRNAs can function through various mechanisms, which can vary depending on the cell type and context.

Despite the annotation of thousands of lncRNAs, our understanding of how the lncRNA sequence influences its function remains nascent. Advances in gene editing and molecular biology technologies have facilitated the identification of functional lncRNAs and their partners [8].

Like mRNAs, many lncRNAs have introns, can undergo post-transcriptional capping and polyadenylation, and are transcribed by RNA polymerase II. However, while our understanding of mRNA biology allows us to predict mRNA function based on their sequence, this is not the case with lncRNAs; little is known about the rules shaping lncRNA sequences and how they relate to their function.

In recent years, the language of lncRNAs has begun to unfold through experimental and computational approaches. In this review, we will delve into these challenges, discuss the current state of knowledge, and outline potential future directions to further unravel the mystery of lncRNA sequences and their functional implications.

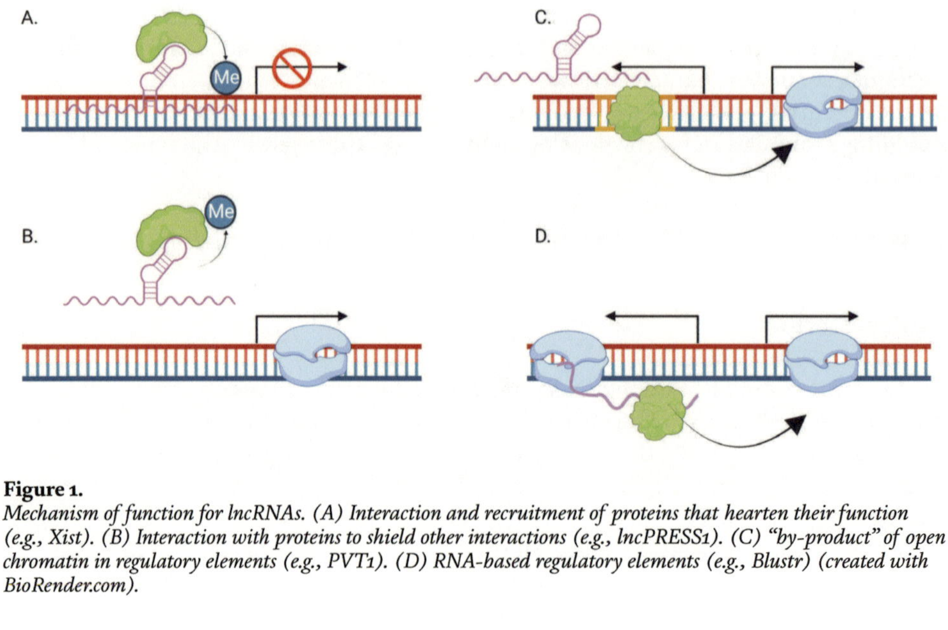

Figure 1.
Mechanism of function for lncRNAs. (A) Interaction and recruitment of proteins that hearten their function (e.g., Xist). (B) Interaction with proteins to shield other interactions (e.g., lncPRESS1). (C) "by-product" of open chromatin in regulatory elements (e.g., PVT1). (D) RNA-based regulatory elements (e.g., Blustr) (created with BioRender.com).

2. Sequence-based classification of lncRNAs and mRNAs

lncRNAs and mRNAs share several characteristics, including the presence of exons and introns, and the processes of splicing, capping, and polyadenylation. Interestingly, lncRNAs have been hypothesized to be infant mRNAs that have yet to gain their coding potential [9]. However, the factors that shape their sequences differ significantly.

The sequences of protein-coding genes (PCGs) primarily serve to hold information in their open-reading frames (ORFs), which is influenced by the frequency of specific codons. PCGs often contain conserved and modular functional domains that can be duplicated across multiple transcripts, resulting in stronger sequence constraints for PCGs compared to lncRNAs.

Several *in silico* approaches have been developed to distinguish between PCGs and lncRNAs based on transcript sequences. These methods can also provide insights into the forces that shape lncRNA sequences. For instance, a high frequency of the GC dinucleotide is a characteristic of higher organisms and is associated with effective splicing [10]. lncRNAs, which typically have fewer exons and weaker splicing efficiency, exhibit lower GC content, making GC dinucleotide composition a useful classifier for distinguishing between lncRNAs and PCGs [11].

Conversely, the TA dinucleotide is strictly regulated in PCGs due to its association with stop codons (TAG, TAA, and TGA), contributing to a higher GC content in PCGs. While the frequency of CG dinucleotides across the genome is low due to their unique regulatory role in DNA methylation, CGs are more enriched in PCGs compared to lncRNAs [12].

Additional sequence-based strategies employ short (2–8 nt) sequential fragments (k-mers) to differentiate between lncRNAs and PCGs. Differences in k-mer frequency between transcript groups may represent functional domains in PCGs or lncRNAs, but they may also mirror single-nucleotide preferences [13–15].

However, PCG transcripts also include significant untranslated regions (UTRs) flanking the ORF. These UTRs, similar to lncRNAs, can interact with RNA-binding proteins (RBPs) and microRNAs (miRNAs). Additionally, many lncRNAs have the potential to encode short ORFs, complicating the classification between PCGs and lncRNAs.

Interestingly, a subset of lncRNAs has been found to harbor short ORFs that can be translated into functional peptides. These lncRNAs, often referred to as "micropeptide-encoding RNAs" or "small peptide-encoding RNAs," challenge the traditional definition of lncRNAs and add another layer of complexity to our understanding of the non-coding transcriptome [16, 17].

Therefore, unexpectedly, ORFs may appear as an important feature of lncRNAs, adding a new dimension to our understanding of these complex molecules.

3. lncRNAs with functional low-complexity sequences

Recent studies have provided valuable insights into the functional roles of lncRNAs that possess low-complexity sequences. Through high-throughput perturbation experiments and correlation analyses of gene expressions, it has been consistently demonstrated that the majority of functional lncRNAs act in a cis-regulatory manner, exerting their regulatory effects on nearby protein-coding genes [18–21]. While multiple models have been proposed to explain this regulation, it is evident that interactions between lncRNAs and mRNAs, chromatin, or proteins often rely on low-sequence specificity.

One prominent function of lncRNAs is their involvement in epigenetic regulation and chromatin accessibility. Numerous studies have identified lncRNAs that interact with the Polycomb repressive complex 2 (PRC2), influencing its activity on nearby genes [22]. PRC2, a core complex involved in histone 3 lysine 27 methylation (H3K27me3), exhibits unique DNA interactions in Drosophila through specific Polycomb response elements (PREs). However, similar motifs have not been identified in mammalian DNA, indicating that PRC2 recruitment to chromatin involves interactions with other mediators. Notably, well-characterized lncRNAs such as X-inactive-specific transcript (XIST) and HOTAIR have been found to interact with PRC2, suggesting their crucial role in PRC2 localization [23–25]. XIST, a conserved lncRNA, consists of six domains (A-F) with tandem repeats. The GC-rich B and E repeats have been identified as essential for PRC2 recruitment and subsequent X chromosome silencing [24].

PRC2 exhibits differential interactions with the four ribonucleotides, displaying a higher affinity for G-rich sequences [26]. Although a clear RNA-binding motif for PRC2 remains elusive, advancements in techniques such as denaturing crosslinking and immunoprecipitation (dCLIP) have enabled the identification of more specific RNA-binding motifs within PRC2 [27]. However, it is important to note that current evidence suggests the existence of partial non-specific interactions between PRC2 and lncRNAs. Functionally, lncRNAs guide PRC2 to specific chromatin regions, facilitating epigenetic changes both in cis and trans. Cis interactions between PRC2 and lncRNAs act as checkpoints for chromatin silencing, while lncRNA-PRC2 interactions may function as decoys, preventing the transcriptional silencing of active genes and reducing the catalytic function of PRC2 [28].

Another model describing interactions with low-sequence specificity involves the recruitment of transcription factors (TFs) during lncRNA metabolism. In such cases, the function of an lncRNA depends on the transcription and processing of its first intron, irrespective of the sequence of exons and introns within the gene.

Notably, studies have demonstrated the essential role of splicing the first intron of the mouse lncRNA 1319, also known as the bivalent locus (Sfmbt2) or Blustr, in regulating the transcription of the protein-coding gene Sfmbt2 [3]. Hence, the metabolism of Blustr governs the regulation of Sfmbt2.

Moreover, research has highlighted the significance of splicing and the recruitment of the spliceosome in the regulation of enhancers, which may overlap with lncRNAs [29, 30]. These findings suggest that many lncRNAs contribute to the regulation of enhancer activity. However, it is crucial to acknowledge that the function of an lncRNA can vary depending on the specific context.

Overall, the functional roles of lncRNAs with low-complexity sequences encompass diverse mechanisms, including the cis-regulation of nearby genes, interactions with PRC2 and other proteins, recruitment of transcription factors, and involvement in enhancer regulation. These roles underscore the complex and multifaceted nature of lncRNA functions in gene expression and epigenetic regulation.

Intriguingly, recent research has demonstrated the formation of 3D membraneless structures called "high-concentration territories" through interactions involving lncRNAs, DNA, and proteins. These interactions, which do not rely on stringent sequence recognition, significantly shape the local environment surrounding lncRNA transcription sites and contribute to the regulation of neighboring gene expressions [31]. This model is particularly attractive in explaining the function of lncRNAs in shaping the local chromatin environment, as they interact with transcription factors and chromatin remodeling agents based on general sequence perforations.

4. lncRNA localization motifs

Long non-coding RNAs (lncRNAs) and messenger RNAs (mRNAs) share several common features. Both are transcribed by RNA polymerase II, undergo splicing, and are often modified with 5′ caps and poly(A) tails. However, their functional locations differ. While mRNAs operate in the cytoplasm and are rapidly exported from the nucleus, many lncRNAs function within the nucleus, necessitating their retention and inefficient export to the cytoplasm. The sequences affecting lncRNA localization have been a major focus of research in the recent years.

The export of RNA molecules longer than 200 nucleotides, including mRNA and lncRNAs, is a tightly regulated process that relies on the formation of a messenger ribonucleoprotein (mRNP) complex. This mRNP formation is intertwined with mRNA/lncRNA processing and is contingent upon its success.

Two key components of the transport machinery are the transcription-and-export complex (TREX) and the nuclear transcription factor X-Box Binding 1 (NFX1). A recent study delineated two distinct RNA export pathways, each favoring different transcript sequences and characteristics. TREX is involved in the export of multi-exonic, C/G-rich transcripts, while NFX1 interacts with A/U-rich transcripts that have fewer and longer exons, with focal high C/G segments. Moreover, NXF1 typically interacts with m6A methylated RNA sites. Both export pathways can target mRNAs and lncRNAs. However, given that lncRNA transcripts tend to be shorter, with fewer exons and lower C/G content, NXF1 is suggested to be the more dominant pathway for lncRNA export [32].

RNA processing is intricately linked with its export to the cytoplasm. Studies have shown that overall splicing efficiency is a strong predictor of RNA localization for both lncRNAs and mRNAs [33, 34]. Key splicing proteins, such as UAP56, interact with nuclear export complexes to enhance export efficiency. However, splicing is not a prerequisite for RNA export [35]. For instance, single-exon transcripts like NORAD are enriched in the cytoplasm, indicating that endogenous intron-less transcripts can be efficiently exported [35].

For transcripts with introns, proper RNA splicing is critical for their stability and export. Recent findings suggest that lncRNA processing is less efficient than that of mRNA due to the extensive use of non-canonical splicing motifs, such as GC-AG, which are associated with less competent splicing compared to the canonical GT-AG motifs [36, 37]. Consequently, lncRNAs tend to be less stable and fail to accumulate in the cytoplasm. However, intron retention can also generate stable and functional lncRNAs. For example, in the case of lncRNA TUG1, intron retention has led to the nuclear enrichment of otherwise cytoplasmic lncRNAs [38]. Therefore, the use of non-canonical and weak splicing motifs may affect lncRNA stability and nuclear export, thereby altering their localization.

The use of non-canonical splicing sites may also influence the polyadenylation process. All mRNA molecules, and most lncRNAs, undergo polyadenylation. This process involves the identification of an A[A/U]UAAA hexamer and a GU/U-rich sequence located approximately 20 nucleotides downstream, followed by RNA cleavage and non-template-dependent synthesis of a poly(A) tail [39]. Interestingly, alternative sequences may also participate in the polyadenylation of mRNAs and lncRNAs [40]. Similar to splicing, polyadenylation occurs during transcription and plays a pivotal role in RNA stability and localization. A connection between polyadenylation and TREX has been reported, with TREX mutations leading to the accumulation of bulk poly(A) RNAs [41]. As most stable lncRNA transcripts are polyadenylated, they are subject to the export machinery.

Two of the most extensively studied lncRNAs, MALAT1 and NEAT1_2, exhibit a unique sequence involved in lncRNA processing. These single-exon lncRNAs are processed by the ribonuclease, RNase P, which is known for its role in tRNA maturation by cleaving the leader sequence at the 5′-end of pre-tRNA. Both MALAT1 and NEAT1_2 generate tRNA-like structures at their 3`-ends that are cleaved by RNase P [42]. However, the unprocessed ends of these transcripts could lead to their rapid degradation by a 3′-to-5′ exonuclease. To protect their 3`-ends, MALAT1 and NEAT1_2 generate a triple-helix structure, composed of two U-rich motifs and an A-rich tract [43]. Currently, this remains a unique sequence feature of these two human lncRNAs, as no other lncRNA has been found to undergo RNase P processing or to have such a unique triple-helix structure. However, structures resembling this triple-helix have been identified in lncRNAs of other genomes, such as the polyadenylated nuclear (PAN) RNA, produced by Kaposi's sarcoma-associated herpesvirus (KSHV). Hence, MALAT1 and NEAT1 represent unique structures of lncRNAs that favor their nuclear retention. However, while some lncRNAs are enriched in the nucleus due to inefficient processing or unique features, other lncRNAs contain specific sequences that actively retain them in the nucleus.

Comprehensive studies using computational and molecular tools have identified additional motifs that influence lncRNA localization. Using a reporter transcript, which by default exports to the cytoplasm, these studies cloned fragments of nuclear-enriched lncRNAs (and mRNAs). Using thousands of gene fragments, these studies identified sequences affecting nuclear localization. These include a fragment of 42 nt Alu [44]. Further analysis refined the motif and identified a core sequence of GCCUCCC. More generally, these studies identify a general tendency toward C-rich sequences that interact with nuclear proteins, such as hnRNPK, SLTM, and SNRNP70 [44, 45]. Nevertheless, both studies highlight the finding that the RNA localization signal is a combination of multiple sequences and is context-dependent. Other high-throughput studies identified the CAGGUGAGU motif, which interacts with U1 snRNA [46], as well as the importance of other transposable elements in transcript localization [47, 48]. These experimental findings underscore the complexity of RNA localization but also pave the way for a better understanding of the relationship between RNA sequence and function.

5. Motif-based lncRNA-protein interactions

While some lncRNA-protein interactions exhibit low-sequence specificity, other RNA-protein complexes rely on interactions between RNA-binding proteins (RBPs) and a sequence-specific motif. RBPs play a critical role in the RNA lifecycle by regulating their processing, localization, stability, and function. When lncRNA was discovered, it was suggested that these transcripts function as scaffolds for RBPs [49]. RBPs' recognition of motifs of 4–6 nt long is done with a certain flexibility and is context-dependent (**Figure 2**).

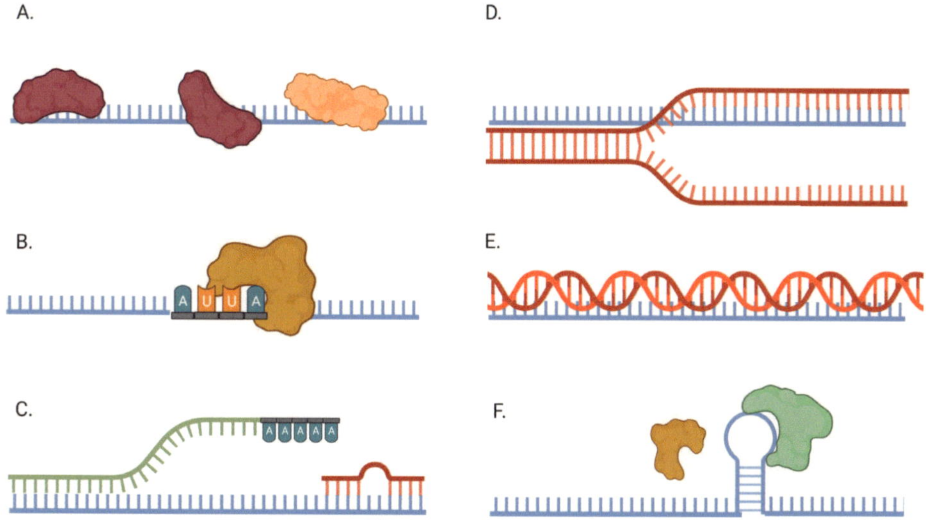

Figure 2.
lncRNAs can interact with different pesters. (A) Non-sequence-specific interactions with proteins. (B) Interaction with RBPs based on motifs. (C) Interactions with mRNA or miRNA. (D) Interaction with DNA (R-loop) (E) interaction with DNA (DNA:DNA:RNA triplex). (F) Self-interactions to generate secondary structures affecting interaction with other molecules. (created with BioRender.com).

The identification of protein-RNA interactions can be performed through RNA pulldown, followed by immunoblotting or mass spectrometry analysis. Alternatively, the pulldown of a suspected protein, followed by quantitative polymerase chain reaction (qPCR) or next-generation sequencing (NGS), can identify which RNA molecules interact with a known protein of interest. However, identifying these interactions is not trivial. The low expression of most lncRNAs, and the vulnerability of RNA molecules to degradation, limits the sensitivity of *in situ* pulldown approaches. On the other hand, *in vitro* approaches may identify non-physiological interactions because of inconsistencies in stoichiometry, localization, and competing interactions. Furthermore, as mentioned above, some proteins are "sticky," and they do not have specific binding sites, but rather interact with sequence or structural motifs on the transcript [50, 51]. While these interactions may be functional, they may also result in misleading interpretations.

NORAD is a highly abundant, cytoplasmatic-enriched, conserved, single-exon lncRNA. Functionally, at the cellular level, NORAD knockdown/−out leads to chromosomal instability and to aging-related phenotypes in mouse models. Pulldown experiments using biotinylated NORAD fragments and cell extract identified PUMILIO homolog 2 (PUM2) as the most enriched interacting protein. These findings were further supported by a sequence analysis that identified at least 15 conserved PUMILIO response elements (PREs) [52]. Further studies demonstrated that NORAD can generate membrane-less bodies in the cytoplasm. This RNA-protein structure generates non-conserved protein-protein interactions that can sponge PUMILIO by overcoming naive protein-RNA stoichiometry. However, NORAD interacts with many other proteins; for example, the RBP SAM68 interacts functionally with UAAA motifs in NORAD [53]. Interestingly, a different pulldown approach, based on *in situ* crosslinking, led to a different list of NORAD-interacting proteins. This analysis highlights the role of nuclear NORAD in DNA replication and genome stability through its direct interaction with the RNA-binding motif protein X-linked (RBMX) [54].

Surprisingly, even though RBPs play a critical role in lncRNA function, only a few specific interaction motifs have been identified. The in-depth characterization of the specific sequences involved in these interactions will provide a further understanding of lncRNA function.

6. The role of secondary structures in lncRNA interactions

Single-stranded RNA (ssRNA) is energetically unstable, folding into secondary and tertiary structures that play a pivotal role in the function of many lncRNAs. These structures generate new binding sites or prevent interactions by altering the context or the availability of motifs. RNA secondary structures are primarily dictated by Watson–Crick base pairing, but non-canonical base pairing is also prevalent, thereby increasing the functional repertoire of RNA [55–57].

An early example of the significance of secondary RNA structures is the lncRNA maternally expressed gene 3 (MEG3), a tumor suppressor that regulates the expressions of P53-downstream genes. Structure and functional analyses identified three structural motifs on MEG3, where the deletion of two affected its function. Interestingly, the function of MEG3 can be restored simply by replacing these deleted motifs with a hybrid RNA of different sequences that can generate the same structures. Therefore, in this case, the structure, rather than the sequence itself, is critical for RNA function [58].

Another example is the lncRNA steroid receptor RNA activator (SRA), which interacts with chromatin modifiers and plays a key role in regulating the expressions of hundreds of genes. Interestingly, SRA is assembled by four domains, whose secondary structure is conserved across 45 vertebrates. However, the sequence itself is much less conserved, indicating that the secondary structure is the driving force in SRA function [59].

Recent studies have shown that dynamic changes in secondary structures can regulate the availability of RNA-binding motifs to their RNA-binding proteins (RBPs). Most motifs are short; therefore, they can be randomly distributed across transcripts. However, higher-order structures affect the availability and affinity of such motifs in protein interactions. It was shown that in a synthetic stem-loop structure, the human U1A (SNRPA) protein interacts better when its motif is located on the loop rather than on the stem part [60]. This way, certain sequences are protected from unwanted interactions. However, the RNA secondary structure is dynamic. Using *in situ* hybridization followed by high-resolution structure analysis, it was shown that the secondary structure of the lncRNA non-coding RNA activated by DNA damage (NORAD) can change under stressful conditions, affecting its biological function [61]. This suggests a new level of lncRNA regulation.

RNA secondary structures may interact directly with DNA-binding proteins. While RBPs usually recognize and interact with single-strand RNA (ssRNA), DNA-binding proteins interact with double-strand sequences. Furthermore, lncRNA secondary structures can function as decoys or activators of transcription factors (TFs) and DNA damage response proteins that identify double-strand DNA. For example, TFs interact with specific DNA sequences that have different chemical/physical characteristics. However, it was shown that some lncRNA sequences can generate double-strand RNA, which can interact with TFs. LncRNA growth arrest-specific 5 (Gas5) can generate a stem-loop structure that mimics the DNA glucocorticoid response elements found in promoters of its target genes. Under physiological conditions of nutrient deprivation, the Gas5 expression is elevated, and it competes with the endogenous sites of glucocorticoid in the promoters to regulate the expressions of metabolic and apoptotic genes [62, 63].

Recent research has further underscored the significance of RNA secondary structures in the regulation of chromatin-associated proteins. A notable example is the interaction between DNMT1, a DNA methyltransferase, and lncRNAs. Multiple studies have shown that DNMT1 can interact with lncRNAs to regulate the expression of specific genes; however, the mode of interaction was not fully clear [64–66]. Recently, it was shown that DNMT1 exhibits a strong and specific affinity for GU-rich RNAs that form a pUG-fold, a noncanonical G-quadruplex [67]. This interaction is not merely incidental; pUG-fold-capable RNAs were found to inhibit DNMT1 activity by preventing the binding of hemimethylated DNA. This suggests that the secondary structure of RNA can directly influence the activity of chromatin-associated proteins, further emphasizing the critical role of RNA secondary structures in the regulation of protein function and activity.

More recently, it has been discovered that besides interacting with DNA, the DNA damage response proteins Ku70/80 can also interact with the secondary structures of lncRNAs. These interactions have been shown to be functional, increasing the activity of such proteins and the efficiency of the DNA damage response [68, 69]. lncRNAs can also interact with TFs and regulate their function. The secondary structure of the mouse lncRNA Braveheart generates a stable stem-loop structure, protecting ssRNA and functioning as a binding site for the nucleic acid binding protein (CNBP/ZNF9) TF [70]. However, it has been recently revealed that these TFs might not interact with lncRNAs through their canonical DNA-binding domain. The TF signal transducer and activator of transcription 3 (STAT3) was found to interact directly with the lncRNA NORAD. However, this interaction is not mediated by the STAT3 DNA-binding domain but rather by aptamer-like interactions with the SH2 domain. This suggests that NORAD does not function as a decoy but rather as a scaffold for the proper function and localization of STAT3 [50].

In conclusion, the higher-order structures of lncRNAs play a crucial role in their function. Furthermore, these secondary structures have recently emerged as promising therapeutic targets for inhibitory small molecules [71]. However, the methods for studying RNA secondary structures are still in their early stages. As the field continues to develop, we anticipate shedding lighter on the important aspects of lncRNA biology and its therapeutic potential.

7. lncRNA interaction with DNA (R-loop) and DNA:DNA:RNA triplexes

R-loops are RNA:DNA hybrids that can form during transcription when the newly synthesized RNA interacts with the DNA template (cis). R-loops play key roles in replication, transcription, and genome integrity. They are enriched near the promoters/transcription start sites (TSSs) of highly expressed genes and near transcription termination sites, suggesting that R-loops play a functional role in transcription regulation [72].

While theoretically, RNA:DNA hybrids can be generated simply by displacement and Watson-Crick hybridization during transcription, R-loop formation is regulated by proteins known to be involved in homologous recombination during DNA damage repair [73]. Importantly, this active R-loop formation mechanism can lead to the hybridization of RNA molecules that share complementary sequences but are transcribed from spatially distinct sites (trans). However, R-loops that are adverse byproducts of transcription must be resolved, typically by RNase H or the Fanconi anemia (FA) pathway; otherwise, these R-loops may interfere with transcription and DNA replication (**Figure 2D**).

Antisense lncRNAs can form R-loops with their sense protein-coding genes (PCGs) and can regulate their expression in cis. For instance, lncRNA ANRASSF1 is transcribed in an antisense orientation to the PCG RASSF1. ANRASSF1 transcription results in the formation of an R-loop, which recruits PRC2 and promotes the epigenetic silencing of RASSF1. Conversely, other R-loop forming lncRNAs are associated with increased transcription. The lncRNA VIM-AS1 is a divergent transcript with PCG vimentin (VIM). The transcription of PCG VIM is regulated by the methylation of a CpG island, and it has been shown that the antisense transcription overlapping the TSS generates an R-loop that inhibits the interaction of NF-kB [74]. Genome-wide studies show that R-loops frequently form at sites of antisense transcription, close to the promoter and TSS, especially at sites enriched in C/G nucleotides. These interactions correlate with low DNA methylation and active promoters [75].

Another significant type of lncRNA interaction with DNA is mediated by the generation of triplexes (DNA:DNA:RNA) between pyrimidine (T and C)-rich RNAs and purine (A and G)-rich DNA *via* Hoogsteen-type hydrogen bonding (**Figure 2E**). The mouse lncRNA Fendrr recruits the epigenetic modulators PRC2 and TrxG/MLL to the promoters of developmental genes. It has been shown that Fendrr interacts with the promoters of the Foxf1 gene in cis and Pitx2 in trans. *In vitro*, these interactions are sensitive to digestion by RNase V1 but not RNase H, which specifically cleaves the DNA:RNA duplex, indicating that Fendrr is involved in triplex (DNA:DNA:RNA) interactions [76]. Another example is lncRNA KHPS1, which is transcribed in antisense orientation to enhancer RNA (eRNA) SPHK1 and generates a DNA:RNA triplex to regulate the expression of SPHK1 mRNA. Interestingly, the incorporation of 7-deaza-purine nucleotides to the triplex-forming region (TFR), which prevents the formation of Hoogsteen base pairing and DNA:RNA triplex formation, abolishes the function of the RNA. Furthermore, by replacing the TFR of KHPS1 with MEG3 [77] and generating a chimeric lncRNA, it was shown that TFRs interact with their RNA partners in an interchangeable manner [78].

Moreover, an analysis of lncRNA-chromatin interactions, based on experimental data, suggests that lncRNAs have cis and trans targets, and lncRNAs regulate multiple genes that can be clustered into specific pathways [77, 79, 80]. Genome-wide analysis indicates that the DNA:RNA triplex is a major mechanism of lncRNA function *in vivo*. An interesting finding based on these analyses is the mechanistic regulatory function of lncRNA HIF1α-AS1 at multiple genomic sites by recruiting the silencing complex under specific physiological contexts [81]. In addition, computational modeling based on chromatin fractionation and an analysis of the dynamic of the dissociation of transcripts from their transcription sites show that lncRNAs overlapping enhancer regions tend to harbor longer before releasing to the nucleoplasm [82]. This may indicate their role in chromatin regulation, directly coupled with their transcription through DNA:RNA interactions. These findings pave the way for a better understanding of these DNA:RNA interactions.

In summary, R-loops and DNA:RNA triplexes constitute important regulatory mechanisms for transcription initiation and termination. In many cases, this is mediated by cis-antisense transcription, which generates unstable RNA. However, accumulating evidence indicates that bona fide lncRNAs can generate R-loops and a DNA:RNA triplex in cis and trans to affect chromatin modification and accessibility.

8. lncRNA interaction with miRNA

MicroRNAs (miRNAs) constitute a significant class of non-coding RNAs (ncRNAs) that modulate post-transcriptional regulation by fine-tuning mRNA translation and stability. miRNAs interact with the miRNA-induced silencing complex (miRISC) and guide the miRISC to its miRNA recognition element (MRE) target based on Watson–Crick base pairing. Notably, in animal cells, most miRNA:MRE interactions are incomplete and include multiple mismatches, which makes the prediction of physiological miRNA targets a challenging task due to the vast number of potential interactions.

Experimental data suggest that miRNAs primarily target mRNA at their 3′ untranslated regions (UTRs) and that miRISC predominantly functions in the cytoplasm, although several studies have demonstrated nuclear function under specific physiological conditions [83, 84]. miRNAs can target mRNAs, as well as lncRNAs, pseudogene-derived transcripts, and circular RNA [85]. As a result, all transcripts compete for the same pool of miRISCs. Consequently, changes in the level of a specific transcript may affect the availability of miRNAs for other targets. This mechanism is described as sponging or competitive endogenous RNA (ceRNA), and it is suggested as a mechanism of action for numerous lncRNAs [86].

While the sponging/ceRNA concept provides an appealing framework to predict lncRNA:mRNA post-transcriptional regulation, accurate prediction is challenging due to the complexity of the synergistic networks and the lack of physiological data required to establish reliable variables and constants [87–91]. Therefore, while growing evidence indicates the attractiveness of sponging as a mechanism and computational predictions suggest a unified language that can explain the function of many lncRNAs, the prediction and validation of specific ceRNA cross-talk remain challenging.

9. Modular sequences in lncRNA and the role of repeat elements

Protein-coding genes (PCGs) are characterized by functional domains, which are modular units with distinct structures and functions [92, 93]. The question of whether long non-coding RNAs (lncRNAs) are similarly structured by modular sequences remains largely unexplored. Transposable elements (TEs), which are enriched in lncRNAs, are proposed to serve as functional domains (**Figure** 3) [93].

TEs are sequences with the ability to duplicate (retroviral oriented) or relocate (non-retroviral elements) within the genome. They are abundant in the human genome, with certain types of TEs specifically enriched in lncRNAs [94]. Previously regarded as parasitic, non-functional elements, TEs are now acknowledged as vital

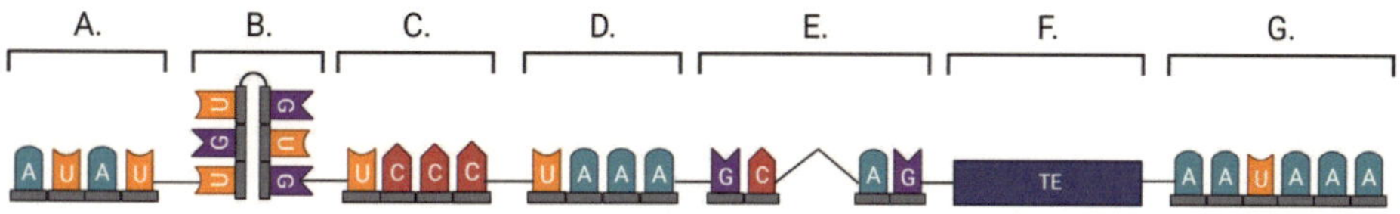

Figure 3.
Sequence motifs in lncRNAs. (A) lncRNA tends to be enriched with a/U. (B) Simple repeats and microsatellites. (C) C-rich sequences involved in nuclear retention. (D) Recognition motifs for proteins and miRNAs. (E) Wick spicing signal. (F) Transposon elements. (G) Polyadenylation and other mRNA motifs. (created with BioRender. com).

to the architecture and function of the genome, driving evolution through their high mobility and consequent rapid genomic sequence changes.

It is estimated that around 80% of human lncRNAs incorporate at least one TE sequence, with approximately 30% of lncRNA sequences being TE-related. TE sequences contribute to the architecture and function of lncRNAs by providing splicing sites, binding domains for RNA-binding proteins (RBPs), and domains for RNA:RNA and RNA:DNA interactions. The insertion of Alu TEs can promote the formation of new exons (a process known as exonization). TEs also contribute sequences for poly(A) signals, as well as sequences affecting lncRNA transcription and localization. The high content of TEs in lncRNAs can promote secondary structures (inverted repeats) that are substrates for A-to-I RNA editing.

As previously discussed, the SIRLOIN motif found in Alu TE elements [44, 95] and sequences from the L1PA16, L2b, MIRb, and MIRc TEs [47] play a role in promoting RNA nuclear retention and the formation of nuclear bodies. Interestingly, when a SINE TE from murine Malat1 was removed using CRISPR-Cas9, lncRNAs were dispersed from the nuclear speckles to the cytoplasm, suggesting a role for TE in its localization and organization. While the SINEB1 TE element is murine-specific, functional sequences can be found in human MALAT1, arguably through convergent evolution [96]. Similarly, the deletion of TE sequences within NEAT1 led to abnormal speckle organization and function [97].

Repetitive elements can also lead to lncRNA-mRNA duplexing. Evidence suggests that these interactions can be functional in mRNA translation and stability [98–102]. However, these interactions are limited by secondary structures, transcript abundance, and localization.

Overall, these repeats play a functional role by interacting with RBPs, RNA, and chromatin, and they demonstrate a proposed mechanism for lncRNA evolution that is based on TEs as functional domains of lncRNAs, termed repeat insertion domains of LncRNAs (RIDLs) [93].

While many lncRNAs contain TEs, which are generated by ancient independent elements, other repetitive sequences are simpler in their nature. Tandem repeats (TRs) and local repeats (LRs) are low-complexity sequences. While TRs appear in multiple loci across the genome, LRs are generally found in a specific locus. Although these sequences are frequent in introns and intergenic DNA, some are located in exons. LRs are enriched in lncRNAs compared to PCGs, and they are suggested to be involved in interactions with RBPs (**Figure 3**). The exonic repeating RNA domain (RRD) in lncRNA functional intergenic repeating RNA element (FIRRE) is a 156-bp repeating sequence that was shown to interact with hnRNPU and to play a critical role in the nuclear retention of the transcript [103].

In addition, lncRNA sequence analyses identified an enrichment of short tandem repeat sequences that can interact with RBPs. For example, pyrimidine-rich noncoding transcript (PNCTR) is a 10-Kb-long lncRNA transcribed by RNA pol I. PNCTR is composed of more than two thousand PTBP1-binding motifs. PNCRT interacts with PRBP1 to generate nuclear bodies called perinucleolar compartments (PNCs). The sequestration and antagonization of PRBP1 by PNCRT play a functional role in splicing regulation [104].

In conclusion, repeat elements, whether they are TEs, TRs, or LRs, play a significant role in the structure and function of lncRNAs. They contribute to the interaction with RBPs, RNA, and chromatin, and they provide a mechanism for lncRNA evolution. These findings pave the way for a better understanding of the role of repeat elements in lncRNAs.

10. Unraveling lncRNA function through sequence analysis

RNA sequences carry the blueprint of their function. While experimental approaches often focus on function first and subsequently search for related sequences, computational approaches primarily compare sequences to identify the functional motifs.

The assumption that functional elements within a sequence are modular allows for the evaluation of lncRNA sequence similarity not merely based on linear alignment, but rather on the accumulation of shared motifs. The linear sequence can be deconstructed into constituent short sequences (K-mers). The distance between any two transcripts can then be described by the correlation between their K-mers, an approach known as Sequence Evaluation from K-mers Representation (SEEKR) [105]. This method has successfully clustered together lncRNAs with known similar functions and differentiated between activating and repressing lncRNAs. Interestingly, SEEKR analysis found that repressor lncRNAs, such as XIST, were enriched in G/C K-mers, while activator lncRNAs, like HOTTIP, clustered together due to the enrichment of A/U nucleotides. Furthermore, K-mer distribution can predict lncRNA subcellular localization, with K-mers enriched in the nucleus found to harbor motifs corresponding with proteins known to affect RNA retention.

Sequence conservation through evolution is a hallmark of functionality. However, many lncRNAs appear to evolve in a species-restricted manner. While most lncRNAs do not show sequence similarity, many reveal syntenic similarities between distinct species. This suggests that lncRNAs maintain their cis-regulatory function without obvious sequence similarities. Furthermore, lncRNAs can change their linear sequence but maintain similar structure and function. While SEEKR looks for the enrichment of multiple K-mers in a non-consecutive order, an analysis of lncRNAs, such as NORAD, identified short RBP recognition motifs in which not only the sequence was conserved, but the sequential order was also maintained through evolution [53]. The LncLOOM framework predicts lncRNA functional motifs based on the assumptions of short motif conservation and motif order.

Collectively, advancements in RNA biology have enabled the development of analytical approaches dedicated to the analysis of lncRNAs. Yet, fundamental questions remain unanswered. Do lncRNAs share a common syntax? Are lncRNAs developed mainly in the context of a specific locus or function? How do the second and third dimensions of lncRNAs function in the syntax of lncRNAs? How can we divide lncRNAs into functional subclasses based on their sequence? How can we predict the function of lncRNAs based on their sequence? The answers to these questions will not only improve our understanding of lncRNA biology but may also open opportunities for novel therapeutic approaches.

11. Conclusions

Identifying a functional sequence in a specific lncRNA remains a challenging task due to the inherent biological nature of lncRNAs and the lack of robust methods. Many of the functional sequences discussed in this review are short and have low sequence restrictions, and their function is highly context-dependent. RNA-binding protein (RBP) and miRNA recognition elements are relatively short and can be found distributed by "chance" along multiple sequences. Furthermore, lncRNAs are enriched in transposable elements (TEs) and other repeat elements that can theoretically interact with many RNA and chromatin sites. Therefore, the function of a sequence depends not only on the context but also on its subcellular localization and abundance [31, 106, 107]. Moreover, recent studies have shown that lncRNA expression is highly dynamic due to its long transcriptional bursts [108], thereby increasing the spatial-temporal function of lncRNAs.

The advancement in experimental techniques has also paved the way for a better understanding of lncRNA sequence-to-function relationships. High-throughput experiments using an expression system with various putative functional sequences enable the identification of motifs that affect RNA nuclear retention. Additionally, CRISPR-based deletions of endogenous sequences or truncated exogenous constructs help to underline functional domains in lncRNAs, especially when combined with evidence from pulldown and phenotypic experiments.

However, the identification of functional domains through phenotypic screening is limited by the complexity of the experimental design and the interpretation of output phenotypes. Therefore, computational approaches, in conjunction with experimental data, will play a crucial role in our understanding of the language of lncRNAs. Sequence analysis shows that lncRNAs differ from protein-coding genes (PCGs), not only due to the lack of substantial open-reading frame (ORF) potential, but also in their nucleotide composition and repetitive elements [13], indicating that lncRNAs function under unique sequence constraints.

Our understanding of the biology and function of lncRNAs is gradually unfolding. This process goes hand in hand not only with developments in analytic and experimental approaches but also with the acknowledgment of lncRNAs' roles in cell function and genome architecture. The role of repeat elements in the structure and function of lncRNAs, as well as the interactions between lncRNAs and proteins or miRNAs, is emerging as significant areas of study.

As we continue to unravel the mysteries of lncRNAs, we anticipate that these non-coding RNAs will provide novel insights into cellular function and offer new avenues for therapeutic intervention. Future research should focus on overcoming the current challenges in identifying functional sequences in lncRNAs, further exploring the potential applications of lncRNAs, and filling the gaps in our current understanding of lncRNAs. The questions posed at the beginning of this chapter remain partially answered, and further research is needed to fully understand the common syntax, functional subclasses, and predictive function of lncRNAs based on their sequence.

Acknowledgements

This work was funded by the Israeli Science Foundation (ISF; grant no. 2228/19) and the Israel Cancer Association (ICA; grant no. 20230069).

Conflict of interest

The authors declare no conflict of interest.

Author details

Assaf C. Bester
Department of Biology, Technion-Israel Institute of Technology, Haifa, Israel

*Address all correspondence to: bestera@technion.ac.il

References

[1] Frankish A et al. GENCODE reference annotation for the human and mouse genomes. Nucleic Acids Research. 2019;**47**(D1):D766-D773. DOI: 10.1093/nar/gky955

[2] Cho SW et al. Promoter of lncRNA gene PVT1 is a tumor-suppressor DNA boundary element. Cell. 2018;**173**(6):1398-1412.e22. DOI: 10.1016/j.cell.2018.03.068

[3] Engreitz JM et al. Local regulation of gene expression by lncRNA promoters, transcription and splicing. Nature. 2016;**539**(7629):452-455. DOI: 10.1038/nature20149

[4] Gil N, Ulitsky I. Production of spliced long noncoding RNAs specifies regions with increased enhancer activity. Cell Systems. 2018;7(5):537-547.e3. DOI: 10.1016/j.cels.2018.10.009

[5] Ali T, Grote P. Beyond the RNA-dependent function of LncRNA genes. ELife. 2020;**9**:e60583. [Published 2020 Oct 23] DOI: 10.7554/eLife.60583

[6] Ji Z, Song R, Regev A, Struhl K. Many lncRNAs, 5'UTRs, and pseudogenes are translated and some are likely to express functional proteins. eLife. 2015;**4**:e08890. DOI: 10.7554/eLife.08890

[7] Matsumoto A et al. mTORC1 and muscle regeneration are regulated by the LINC00961-encoded SPAR polypeptide. Nature. 2017;**541**(7636):228-232. DOI: 10.1038/nature21034

[8] Hazan J, Bester AC. CRISPR-based approaches for the high-throughput characterization of long non-coding RNAs. Non-coding RNA. 2021;7(4):79. DOI: 10.3390/ncrna7040079

[9] An NA, Zhang J, Mo F, et al. De novo genes with an lncRNA origin encode unique human brain developmental functionality. Nature Ecology and Evolution. 2023;7:264-278. DOI: 10.1038/s41559-022-01925-6

[10] Amit M et al. Differential GC content between exons and introns establishes distinct strategies of splice-site recognition. Cell Reports. 2012;**1**(5):543-556. DOI: 10.1016/j.celrep.2012.03.013

[11] Haerty W, Ponting CP. Unexpected selection to retain high GC content and splicing enhancers within exons of multiexonic lncRNA loci. RNA. 2015;**21**(3):333-346. DOI: 10.1261/rna.047324.114

[12] Ulveling D, Dinger ME, Francastel C, Hubé F. Identification of a dinucleotide signature that discriminates coding from non-coding long RNAs. Frontiers in Genetics. 2014;**5**:316. DOI: 10.3389/fgene.2014.00316

[13] Wen J, Liu Y, Shi Y, Huang H, Deng B, Xiao X. A classification model for lncRNA and mRNA based on k-mers and a convolutional neural network. BMC Bioinformatics. 2019;**20**(1):469. DOI: 10.1186/s12859-019-3039-3

[14] Li A, Zhang J, Zhou Z. PLEK: A tool for predicting long non-coding RNAs and messenger RNAs based on an improved k-mer scheme. BMC Bioinformatics. 2014;**15**:311. DOI: 10.1186/1471-2105-15-311

[15] Hu L, Xu Z, Hu B, Lu ZJ. COME: A robust coding potential calculation tool for lncRNA identification and characterization based on multiple features. Nucleic Acids Research. 2017;**45**(1):e2. DOI: 10.1093/nar/gkw798

[16] Anderson DM et al. A micropeptide encoded by a putative long noncoding

RNA regulates muscle performance. Cell. 2015;**160**(4):595-606. DOI: 10.1016/j.cell.2015.01.009

[17] Wright BW, Yi Z, Weissman JS, Chen J. The dark proteome: translation from noncanonical open reading frames, 2022. doi: 10.1016/j.tcb.2021.10.010.

[18] Bester AC et al. An integrated genome-wide CRISPRa approach to functionalize lncRNAs in drug resistance. Cell. 2018;**173**(3):649-664.e20. DOI: 10.1016/j.cell.2018.03.052

[19] Liu SJ et al. CRISPRi-based genome-scale identification of functional long noncoding RNA loci in human cells. Science. 2017;**355**:eaah7111. DOI: 10.1126/science.aah7111

[20] Joung J et al. Genome-scale CRISPR-Cas9 knockout and transcriptional activation screening. Nature Protocols. 2017;**12**(4):828-863. DOI: 10.1038/nprot.2017.016

[21] Derrien T et al. The GENCODE v7 catalog of human long noncoding RNAs: Analysis of their gene structure, evolution, and expression. Genome Research. 2012;**22**(9):1775-1789. DOI: 10.1101/gr.132159.111

[22] Khalil AM et al. Many human large intergenic noncoding RNAs associate with chromatin-modifying complexes and affect gene expression. Proceedings of the National Academy of Sciences of the United States of America. 2009;**106**(28):11667-11672. DOI: 10.1073/pnas.0904715106

[23] Colognori D, Sunwoo H, Kriz AJ, Wang CY, Lee JT. Xist deletional analysis reveals an interdependency between Xist RNA and polycomb complexes for spreading along the inactive X. Molecular Cell. 2019;**74**(1):101-117.e10. DOI: 10.1016/j.molcel.2019.01.015

[24] Dixon-McDougall T, Brown CJ. Independent domains for recruitment of PRC1 and PRC2 by human XIST. PLoS Genetics. 2021;**17**(3):e1009123. DOI: 10.1371/journal.pgen.1009123

[25] Bousard A et al. The role of Xist-mediated polycomb recruitment in the initiation of X-chromosome inactivation. EMBO Reports. 2019;**20**(10):e48019. DOI: 10.15252/embr.201948019

[26] Wang X et al. Targeting of polycomb repressive complex 2 to RNA by short repeats of consecutive guanines. Molecular Cell. 2017;**65**(6):1056-1067.e5. DOI: 10.1016/j.molcel.2017.02.003

[27] Rosenberg M et al. Motif-driven interactions between RNA and PRC2 are rheostats that regulate transcription elongation. Nature Structural & Molecular Biology. 2021;**28**(1):103-117. DOI: 10.1038/s41594-020-00535-9

[28] Kaneko S, Son J, Bonasio R, Shen SS, Reinberg D. Nascent RNA interaction keeps PRC2 activity poised and in check. Genes & Development. 2014;**28**(18):1983-1988. DOI: 10.1101/gad.247940.114

[29] Gil N, Ulitsky I. Regulation of gene expression by cis-acting long non-coding RNAs. Nature Reviews. Genetics. 2020;**21**(2):102-117. DOI: 10.1038/s41576-019-0184-5

[30] Tan JY, Biasini A, Young RS, Marques AC. Splicing of enhancer-associated lincRNAs contributes to enhancer activity. Life Science Alliance. 2020;**3**(4):1-12. DOI: 10.26508/LSA.202000663

[31] Quinodoz SA et al. RNA promotes the formation of spatial compartments in the nucleus. Cell. 2021;**184**(23):5775-5790.e30. DOI: 10.1016/j.cell.2021.10.014

[32] Zuckerman B, Ron M, Mikl M, Segal E, Ulitsky I. Gene architecture and sequence composition underpin selective dependency of nuclear export of long RNAs on NXF1 and the TREX complex. Molecular Cell. 2020;**79**(2):251-267.e6. DOI: 10.1016/j.molcel.2020.05.013

[33] Wu KE, Parker KR, Fazal FM, Chang HY, Zou J. RNA-GPS predicts high-resolution RNA subcellular localization and highlights the role of splicing. RNA. 2020;**26**(7):851-865. DOI: 10.1261/rna.074161.119

[34] Zuckerman B, Ulitsky I. Predictive models of subcellular localization of long RNAs. RNA. 2019;**25**(5):557-572. DOI: 10.1261/rna.068288.118

[35] Lei H, Dias AP, Reed R. Export and stability of naturally intronless mRNAs require specific coding region sequences and the TREX mRNA export complex. Proceedings of the National Academy of Sciences of the United States of America. 2011;**108**(44):17985-17990. DOI: 10.1073/pnas.1113076108

[36] Krchňáková Z et al. Splicing of long non-coding RNAs primarily depends on polypyrimidine tract and 5′ splice-site sequences due to weak interactions with SR proteins. Nucleic Acids Research. 2019;**47**(2):911-928. DOI: 10.1093/nar/gky1147

[37] Abou Alezz M, Celli L, Belotti G, Lisa A, Bione S. GC-AG introns features in long non-coding and protein-coding genes suggest their role in gene expression regulation. Frontiers in Genetics. 2020;**11**:488. DOI: 10.3389/fgene.2020.00488

[38] Dumbović G et al. Nuclear compartmentalization of TERT mRNA and TUG1 lncRNA is driven by intron retention. Nature Communications. 2021;**12**(1):3308. DOI: 10.1038/s41467-021-23221-w

[39] Richter JD. Cytoplasmic polyadenylation in development and beyond. Microbiology and Molecular Biology Reviews. 1999;**63**(2):446-456. DOI: 10.1128/MMBR.63.2.446-456.1999

[40] Li X-Q, Du D. Motif types, motif locations and base composition patterns around the RNA polyadenylation site in microorganisms, plants and animals. BMC Evolutionary Biology. 2014;**14**:162. DOI: 10.1186/s12862-014-0162-7

[41] Katahira J, Okuzaki D, Inoue H, Yoneda Y, Maehara K, Ohkawa Y. Human TREX component Thoc5 affects alternative polyadenylation site choice by recruiting mammalian cleavage factor I. Nucleic Acids Research. 2013;**41**(14):7060-7072. DOI: 10.1093/nar/gkt414

[42] Wilusz JE, Freier SM, Spector DL. 3′ end processing of a long nuclear-retained noncoding RNA yields a tRNA-like cytoplasmic RNA. Cell. 2008;**135**(5):919-932. DOI: 10.1016/j.cell.2008.10.012

[43] Wilusz JE, JnBaptiste CK, Lu LY, Kuhn C-D, Joshua-Tor L, Sharp PA. A triple helix stabilizes the 3′ ends of long noncoding RNAs that lack poly(a) tails. Genes & Development. 2012;**26**(21):2392-2407. DOI: 10.1101/gad.204438.112

[44] Lubelsky Y, Ulitsky I. Sequences enriched in Alu repeats drive nuclear localization of long RNAs in human cells. Nature. 2018;**555**(7694):107-111. DOI: 10.1038/nature25757

[45] Zhang B,Gunawardane L, Niazi F, Jahanbani F, Chen X,Valadkhan S. A novel RNA motif mediates the strict nuclear localization of a long noncoding RNA. Molecular and Cellular Biology. Jun 2014;**34**(12):2318-2329. DOI: 10.1128/MCB.01673-13

[46] Yin Y et al. U1 snRNP regulates chromatin retention of noncoding RNAs. Nature. 2020;**580**(7801):147-150. DOI: 10.1038/s41586-020-2105-3

[47] Carlevaro-Fita J, Johnson R. Global positioning system: Understanding long noncoding RNAs through subcellular localization. Molecular Cell. 2019;**73**(5):869-883. DOI: 10.1016/j.molcel.2019.02.008

[48] Ron M, Ulitsky I. Context-specific effects of sequence elements on subcellular localization of linear and circular RNAs. Nature Communications. 2022;**13**(1):2481. DOI: 10.1038/s41467-022-30183-0

[49] Poramba-Liyanage DW et al. Inhibition of transcription leads to rewiring of locus-specific chromatin proteomes. Genome Research. 2020;**30**(4):635-646. DOI: 10.1101/gr.256255.119

[50] Dvir S et al. Uncovering the RNA-binding protein landscape in the pluripotency network of human embryonic stem cells. Cell Reports. 2021;**35**(9):109198. DOI: 10.1016/j.celrep.2021.109198

[51] Zhang Z, Sun W, Shi T, Lu P, Zhuang M, Liu JL. Capturing RNA-protein interaction via CRUIS. Nucleic Acids Research. 2020;**48**(9):e52-e52. DOI: 10.1093/nar/gkaa143

[52] Lee S et al. Noncoding RNA NORAD regulates genomic stability by sequestering PUMILIO proteins. Cell. 2016;**164**(1-2):69-80. DOI: 10.1016/j.cell.2015.12.017

[53] Tichon A, Perry RBT, Stojic L, Ulitsky I. SAM68 is required for regulation of pumilio by the NORAD long noncoding RNA. Genes & Development. 2018;**32**(1):70-78. DOI: 10.1101/gad.309138.117

[54] Munschauer M et al. The NORAD lncRNA assembles a topoisomerase complex critical for genome stability. Nature. 2018;**561**(7721):132-136. DOI: 10.1038/s41586-018-0453-z

[55] Yang SY et al. Transcriptome-wide identification of transient RNA G-quadruplexes in human cells. Nature Communications. 2018;**9**(1):4730. DOI: 10.1038/s41467-018-07224-8

[56] Lyu K, Chow EY-C, Mou X, Chan T-F, Kwok CK. RNA G-quadruplexes (rG4s): Genomics and biological functions. Nucleic Acids Research. 2021;**49**(10):5426-5450. DOI: 10.1093/nar/gkab187

[57] Simko EAJ et al. G-quadruplexes offer a conserved structural motif for NONO recruitment to NEAT1 architectural lncRNA. Nucleic Acids Research. 2020;**48**(13):7421-7438. DOI: 10.1093/nar/gkaa475

[58] Zhang X et al. Maternally expressed gene 3 (MEG3) noncoding ribonucleic acid: Isoform structure, expression, and functions. Endocrinology. 2010;**151**(3):939-947. DOI: 10.1210/en.2009-0657

[59] Novikova IV, Hennelly SP, Sanbonmatsu KY. Structural architecture of the human long non-coding RNA, steroid receptor RNA activator. Nucleic Acids Research. 2012;**40**(11):5034-5051. DOI: 10.1093/nar/gks071

[60] Komatsu KR, Taya T, Matsumoto S, Miyashita E, Kashida S, Saito H. RNA structure-wide discovery of functional interactions with multiplexed RNA motif library. Nature Communications. 2020;**11**(1):6275. DOI: 10.1038/s41467-020-19699-5

[61] Ziv O, Farberov S, Lau JY, Miska E, Kudla G, Ulitsky I. Structural features within the NORAD long noncoding RNA underlie efficient repression of Pumilio activity. bioRxiv. 2021. DOI: 10.1101/2021.11.19.469243

[62] Parsonnet NV, Lammer NC, Holmes ZE, Batey RT, Wuttke DS. The glucocorticoid receptor DNA-binding domain recognizes RNA hairpin structures with high affinity. Nucleic Acids Research. 2019;**47**(15):8180-8192. DOI: 10.1093/nar/gkz486

[63] Kino T, Hurt DE, Ichijo T, Nader N, Chrousos GP. Noncoding RNA gas5 is a growth arrest- and starvation-associated repressor of the glucocorticoid receptor. Science Signaling. 2010;**3**(107):ra8. DOI: 10.1126/scisignal.2000568

[64] Somasundaram S et al. The DNMT1-associated lincRNA DACOR1 reprograms genome-wide DNA methylation in colon cancer 06 biological sciences 0604 genetics. Clinical Epigenetics. 2018;**10**(1):1-15. DOI: 10.1186/S13148-018-0555-3/FIGURES/10

[65] Di Ruscio A et al. DNMT1-interacting RNAs block gene-specific DNA methylation. Nature. 2013;**503**(7476): 371-376. DOI: 10.1038/nature12598

[66] Merry CR et al. DNMT1-associated long non-coding RNAs regulate global gene expression and DNA methylation in colon cancer. 2015. doi: 10.1093/hmg/ddv343

[67] Jansson-Fritzberg LI et al. DNMT1 inhibition by pUG-fold quadruplex RNA. RNA. 2023;**29**(3):346-360. DOI: 10.1261/RNA.079479.122/-/DC1

[68] Haemmig S et al. Long noncoding RNA SNHG12 integrates a DNA-PK-mediated DNA damage response and vascular senescence. Science Translational Medicine. 2020;**12**(531):eaaw1868. DOI: 10.1126/scitranslmed.aaw1868

[69] Unfried JP et al. Long noncoding RNA NIHCOLE promotes ligation efficiency of DNA double-Strand breaks in hepatocellular carcinoma. Cancer Research. 2021;**81**(19):4910-4925. DOI: 10.1158/0008-5472.CAN-21-0463

[70] Xue Z et al. A G-rich motif in the lncRNA Braveheart interacts with a zinc-finger transcription factor to specify the cardiovascular lineage. Molecular Cell. 2016;**64**(1):37-50. DOI: 10.1016/j.molcel.2016.08.010

[71] Childs-Disney JL, Yang X, Gibaut QMR, Tong Y, Batey RT, Disney MD. Targeting RNA structures with small molecules. Nature Reviews. Drug Discovery. 2022;**21**(10):736-762. DOI: 10.1038/s41573-022-00521-4

[72] Aguilera A, García-Muse T. R Loops: From transcription Byproducts to threats to genome stability. Molecular Cell. Apr 2012;**46**(2)115-124. DOI: 10.1016/j.molcel.2012.04.009

[73] Wahba L, Gore SK, Koshland D. The homologous recombination machinery modulates the formation of RNA-DNA hybrids and associated chromosome instability. eLife. Jun 2013;**2**:e00505. DOI: 10.7554/eLife.00505

[74] Boque-Sastre R et al. Head-to-head antisense transcription and R-loop formation promotes transcriptional activation. Proceedings of the National Academy of Sciences of the United States of America. 2015;**112**(18):5785-5790. DOI: 10.1073/pnas.1421197112

[75] Sanz LA et al. Prevalent, dynamic, and conserved R-loop structures associate with specific Epigenomic signatures in mammals. Molecular Cell.

2016;**63**(1):167-178. DOI: 10.1016/j.molcel.2016.05.032

[76] Grote P et al. The tissue-specific lncRNA Fendrr is an essential regulator of heart and body wall development in the mouse. Developmental Cell. 2013;**24**(2):206-214. DOI: 10.1016/j.devcel.2012.12.012

[77] Mondal T et al. MEG3 long noncoding RNA regulates the TGF-β pathway genes through formation of RNA-DNA triplex structures. Nature Communications. 2015;**6**:7743. DOI: 10.1038/ncomms8743

[78] Blank-Giwojna A, Postepska-Igielska A, Grummt I. lncRNA KHPS1 activates a poised enhancer by triplex-dependent recruitment of Epigenomic regulators. Cell Reports. 2019;**26**(11):2904-2915.e4. DOI: 10.1016/j.celrep.2019.02.059

[79] Zhang G et al. Comprehensive analysis of long noncoding RNA (lncRNA)-chromatin interactions reveals lncRNA functions dependent on binding diverse regulatory elements. The Journal of Biological Chemistry. 2019;**294**(43):15613-15622. DOI: 10.1074/jbc.RA119.008732

[80] Sentürk Cetin N, Kuo C-C, Ribarska T, Li R, Costa IG, Grummt I. Isolation and genome-wide characterization of cellular DNA:RNA triplex structures. Nucleic Acids Research. 2019;**47**(5):2306-2321. DOI: 10.1093/nar/gky1305

[81] Leisegang MS, Bains JK, Seredinski S, et al. HIF1α-AS1 is a DNA:DNA:RNA triplex-forming lncRNA interacting with the HUSH complex. Nature Communications. 2022;**13**:6563. DOI: 10.1038/s41467-022-34252-2

[82] Lin Y, Li C, Xiong W, Fan L, Pan H, Li Y. ARSD, a novel ERα downstream target gene, inhibits proliferation and migration of breast cancer cells via activating hippo/YAP pathway. Cell Death & Disease. 2021;**12**(11):1042. DOI: 10.1038/s41419-021-04338-8

[83] Castanotto D et al. A stress-induced response complex (SIRC) shuttles miRNAs, siRNAs, and oligonucleotides to the nucleus. Proceedings of the National Academy of Sciences of the United States of America. 2018;**115**(25):E5756-E5765. DOI: 10.1073/pnas.1721346115

[84] La Rocca G, Cavalieri V. Roles of the Core components of the mammalian miRISC in chromatin biology. Genes (Basel). 2022;**13**(3):414. DOI: 10.3390/genes13030414

[85] Hansen TB et al. Natural RNA circles function as efficient microRNA sponges. Nature. 2013;**495**(7441):384-388. DOI: 10.1038/nature11993

[86] Thomson DW, Dinger ME. Endogenous microRNA sponges: Evidence and controversy. Nature Reviews. Genetics. 2016;**17**(5):272-283. DOI: 10.1038/nrg.2016.20

[87] Broderick JA, Zamore PD. Competitive endogenous RNAs cannot alter microRNA function in vivo. Molecular Cell. 2014;**54**(5):711-713. DOI: 10.1016/j.molcel.2014.05.023

[88] Bosson AD, Zamudio JR, Sharp PA. Endogenous miRNA and target concentrations determine susceptibility to potential ceRNA competition. Molecular Cell. 2014;**56**(3):347-359. DOI: 10.1016/j.molcel.2014.09.018

[89] Denzler R, Agarwal V, Stefano J, Bartel DP, Stoffel M. Assessing the ceRNA hypothesis with quantitative measurements of miRNA and target abundance. Molecular Cell.

2014;**54**(5):766-776. DOI: 10.1016/j.molcel.2014.03.045

[90] Denzler R, McGeary SE, Title AC, Agarwal V, Bartel DP, Stoffel M. Impact of MicroRNA levels, target-site complementarity, and cooperativity on competing endogenous RNA-regulated gene expression. Molecular Cell. 2016;**64**(3):565-579. DOI: 10.1016/j.molcel.2016.09.027

[91] Jens M, Rajewsky N. Competition between target sites of regulators shapes post-transcriptional gene regulation. Nature Reviews. Genetics. 2015;**16**(2):113-126. DOI: 10.1038/nrg3853

[92] Guttman M, Rinn JL. Modular regulatory principles of large non-coding RNAs. Nature. 2012;**482**(7385):339-346. DOI: 10.1038/nature10887

[93] Johnson R, Guigó R. The RIDL hypothesis: Transposable elements as functional domains of long noncoding RNAs. RNA. 2014;**20**(7):959-976. DOI: 10.1261/rna.044560.114

[94] Kapusta A et al. Transposable elements are major contributors to the origin, diversification, and regulation of vertebrate long noncoding RNAs. PLoS Genetics. 2013;**9**(4):e1003470. DOI: 10.1371/journal.pgen.1003470

[95] Chillón I, Pyle AM. Inverted repeat Alu elements in the human lincRNA-p21 adopt a conserved secondary structure that regulates RNA function. Nucleic Acids Research. 2016;**44**(19):9462-9471. DOI: 10.1093/nar/gkw599

[96] Nguyen TM et al. The SINEB1 element in the long non-coding RNA Malat1 is necessary for TDP-43 proteostasis. Nucleic Acids Research. 2020;**48**(5):2621-2642. DOI: 10.1093/nar/gkz1176

[97] Yamazaki T et al. Functional domains of NEAT1 architectural lncRNA induce Paraspeckle assembly through phase separation. Molecular Cell. 2018;**70**(6):1038-1053.e7. DOI: 10.1016/j.molcel.2018.05.019

[98] Schein A, Zucchelli S, Kauppinen S, Gustincich S, Carninci P. Identification of antisense long noncoding RNAs that function as SINEUPs in human cells. Scientific Reports. 2016;**6**:33605. DOI: 10.1038/srep33605

[99] Gong C, Maquat LE. lncRNAs transactivate STAU1-mediated mRNA decay by duplexing with 3' UTRs via Alu elements. Nature. 2011;**470**(7333):284-288. DOI: 10.1038/nature09701

[100] Fasolo F et al. The RNA-binding protein ILF3 binds to transposable element sequences in SINEUP lncRNAs. The FASEB Journal. 2019;**33**(12):13572-13589. DOI: 10.1096/fj.201901618RR

[101] Kim YK, Furic L, Parisien M, Major F, DesGroseillers L, Maquat LE. Staufen1 regulates diverse classes of mammalian transcripts. The EMBO Journal. 2007;**26**(11):2670-2681. DOI: 10.1038/sj.emboj.7601712

[102] Kretz M et al. Control of somatic tissue differentiation by the long non-coding RNA TINCR. Nature. 2013;**493**(7431):231-235. DOI: 10.1038/nature11661

[103] Hacisuleyman E, Shukla CJ, Weiner CL, Rinn JL. Function and evolution of local repeats in the Firre locus. Nature Communications. 2016;**7**:11021. DOI: 10.1038/ncomms11021

[104] Yap K, Mukhina S, Zhang G, Tan JSC, Ong HS, Makeyev EV. A short tandem repeat-enriched RNA assembles a nuclear compartment to control

alternative splicing and promote cell survival. Molecular Cell. 2018;**72**(3):525-540.e13. DOI: 10.1016/j.molcel.2018.08.041

[105] Kirk JM et al. Functional classification of long non-coding RNAs by k-mer content. Nature Genetics. 2018;**50**(10):1474-1482. DOI: 10.1038/s41588-018-0207-8

[106] Elguindy MM, Mendell JT. NORAD-induced Pumilio phase separation is required for genome stability. Nature. 2021;**595**(7866):303-308. DOI: 10.1038/s41586-021-03633-w

[107] Cabili MN et al. Localization and abundance analysis of human lncRNAs at single-cell and single-molecule resolution. Genome Biology. 2015;**16**:20. DOI: 10.1186/s13059-015-0586-4

[108] Johnsson P et al. Transcriptional kinetics and molecular functions of long noncoding RNAs. Nature Genetics. 2022;**54**(3):306-317. DOI: 10.1038/s41588-022-01014-1

Chapter 2

Transcribed-Ultra Conserved Regions (T-UCRs) a New Light on a Dark Matter

Maria Radanova

Abstract

Transcribed Ultra-Conserved Regions (T-UCRs) are a novel class of long non-coding RNAs derived from Ultra-Conserved Regions (UCRs) of DNA. The discovery of cancer-specific mutations in UCRs and their location in cancer-associated genomic regions suggests that T-UCRs also play a role in carcinogenesis. However, the mechanisms behind their actions remain unclear. Their interactions with microRNAs are not well understood and are currently a subject of debate. Like other non-coding RNAs, T-UCRs exhibit tissue- and disease-specific expression, making them promising candidates for biomarkers or therapeutic targets in cancer and other diseases. This chapter aims to review the current knowledge on the functional effects of T-UCRs in cancer and other diseases, discuss the role of T-UCRs as regulators and regulated, and present their potential as disease monitoring biomarkers.

Keywords: ultra-conserved regions, UCR, transcribed-ultra conserved regions, T-UCRs, non-coding RNAs

1. Introduction

The genome is a malleable substance. A large portion of it is made up of so-called "dark matter"—non-coding protein genomic sequences. Until recently, the transcripts of this part of the genome were considered unnecessary and non-functional. However, it is now known that they carry important information and play a major role in regulating the phenotype. It has been proven that the "dark matter" of the genome is transcribed into various types of RNA, most of which do not encode proteins, known as non-coding RNAs [1]. Based on transcript length, non-coding RNAs are divided into two major groups. Those longer than 200 nucleotides are referred to as long non-coding RNAs (lncRNAs), while those shorter than 200 nucleotides are known as short non-coding RNAs (sncRNAs). Among the sncRNAs, well-studied microRNAs (miRs) are involved in the process of RNA interference [2]. Transcribed Ultra-Conserved Regions (T-UCRs) belong to the lncRNA group. T-UCRs are transcribed from 481 ultra-conserved DNA sequence regions (UCRs). T-UCRs were discovered in recent years and have quickly become a new field of exploration.

The chapter aims to provide a critical review of the current scientific understanding of the mechanisms of action of T-UCRs in pathological conditions, to present possible mechanisms for regulating their expression and to summarize research that demonstrates the diagnostic, prognostic, and predictive potential of T-UCRs as biomarkers.

2. Discovery and classification of UCRs

UCRs were first described by Bejerano et al. and were found to be present on all chromosomes, except chromosomes 21 and Y, with an uneven distribution [3]. They are longer than 200 base pairs and are functional but do not encode proteins [4]. UCRs are often located in fragile sites and genomic regions associated with tumor-related genes, and 93% of them are transcribed in at least one normal human tissue [5]. They are characterized by absolute conservation, with no insertions or deletions, and are 100% identical in humans, rats, and mice [3].

Their initial classification of UCRs was made by Bejarano et al. with a focus on their intersection with protein-coding regions. The authors divided the 481 UCRs into three groups: 111 were classified as *partly exonic* (or exonic for short), 114 as *possibly exonic* due to inconclusive intersection with a protein-coding sequence, and the remaining 256 were classified as *nonexonic* [3]. The first two groups are randomly distributed along the chromosomes, while the *nonexonic* elements tend to cluster in introns or intergenic regions near transcription factors and developmental genes.

In 2010, Mestdagh et al. proposed a more detailed classification of UCRs. The genomic groups became five: *exonic*, *intronic*, *exon-containing*, *intergenic*, and *partially exonic* UCRs [6]. A large part of UCRs, more than 70% are *intronic* and *intergenic* and most of them are transcriptionally active [5–7]. The UCRs groups correspond to their transcription product T-UCRs (**Figure 1**).

T-UCRs can be divided into two groups: *intragenic* and *intergenic*. The intragenic group includes T-UCRs whose UCRs are located within exons or introns of their host genes, such as *exonic*, *intronic*, or T-UCRs whose UCRs span an exon such as *exon-containing*, and *partly exonic*. Intergenic T-UCRs are those whose UCRs are located outside of current genomic annotations. This separation of T-UCRs is mainly based on the localization of their UCRs and is somewhat conditional. Moreover, many *intronic* UCRs have been discovered in the antisense orientation relative to their host

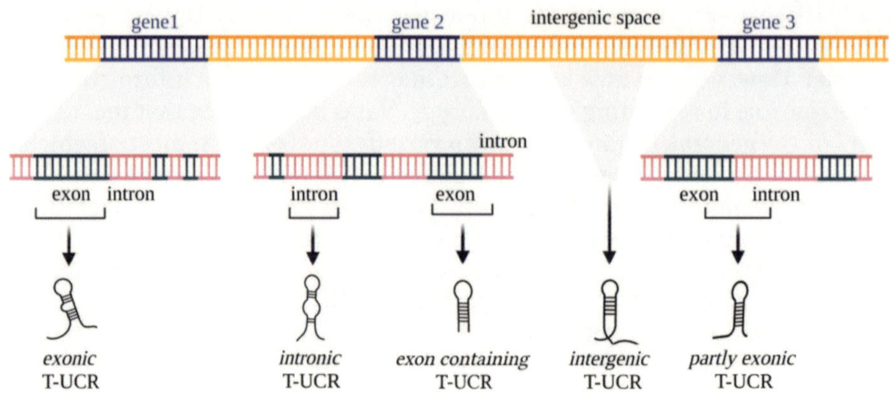

Figure 1.
Different types of T-UCRs according to genomic localization of UCRs. (Figure created with BioRender.com.)

gene. This suggests that their transcripts T-UCRs are not simply products of intronic transcription of known genes, but rather are genuine independent noncoding transcripts [3]. The T-UCRs which correspond to the sense genome sequence, referred to as "+", and those that correspond to the antisense sequence, referred to as "+A". The designation of T-UCRs includes this information, following the "uc." (from ultraconserved) and the number indicating their chromosomal location (for example uc.346+ or uc.479+A).

In recent years, the number of studies on the functions of T-UCRs in the biology of different diseases has rapidly increased. These studies have been summarized and commented on in a number of review articles [4, 7–15]. However, the need for discussions on this topic remains at this stage because what is known about T-UCRs and UCRs is less than what is unknown. It is important to differentiate between the roles of T-UCRs and the UCRs from which they are transcribed. T-UCRs may not represent the entire RNA transcript, and their sequences often overlap with ultraconserved enhancers or are contained within other annotated genes. On the other hand, Fedorova et al. argued that ultra-conserved non-coding elements do not represent non-coding RNAs, as T-UCRs are assumed to be [16]. Although they share common functions with lncRNAs, T-UCRs also exhibit many specific ones. Additionally, they do not quite fit the view of ncRNAs, which primarily have evolutionary conservation in structure rather than in sequence [16]. To fully understand T-UCRs, it is necessary to clarify their structure, the factors that influence their expression, and their effects on both the host and neighboring genes.

3. Mechanisms of regulation of gene expression by T-UCRs

T-UCRs play a crucial role in regulating gene expression by influencing various processes, including pri-miR processing and maturation, mRNA degradation, transcription regulation, epigenetic modulation [7, 17].

Efforts to understand the general mechanisms of action of T-UCRs face challenges due to the fact that some of these mechanisms are also involved in regulating the expression of T-UCRs themselves. It remains unclear whether the pathogenic role of T-UCRs is more closely related to their direct participation in the regulation of the expression of disease-related genes or if the dysregulation of their own expression is the determining factor. This raises an interesting question: how can a T-UCR simultaneously act as both a regulator and a regulated entity?

3.1 Interaction between T-UCRs and mRNA

An established mechanism of action for T-UCRs includes their ability to directly bind to the 3'UTR of mRNAs, leading primarily to the degradation of the gene's mRNA. For example, uc.206 specifically targets mRNA of the pro-apoptotic gene *P53*, promoting tumor cell growth in cervical cancer [18]. Additionally, uc.338 interacts directly within the 3'UTR of TIMP-1 mRNA and negatively regulates it at a post-transcriptional level. Through this mechanism, uc.338 decreases TIMP-1 but increases MMP9 expression in the TIMP-1-MMP9 signaling pathway, which may explain the tumor progression in colorectal and cervical cancers [19, 20]. HSPA12B mRNA is a direct target of uc.454. In this way, uc.454 may induce apoptosis of lung tumor cells by decreasing the expression of HSPA12B and affecting the ratio of Bcl-2/Bax [21]. Uc.454 can also inhibit migration and invasion in lung cancer by interacting with

K-Ras at the posttranscriptional level, affecting the K-Ras/P63/MMP9 pathway [22]. A study by Goto et al. suggests that uc.416+A operates through a similar mechanism on the mRNA of several proteins. Uc.416+A is associated with cell growth through the downregulation of expression of inhibitors of proliferation, differentiation, and migration of tumors, such as IGFBP6, and the upregulation of proteins associated with malignant neoplasms, such as HOXB5 [23]. Mechanistically, uc.189 inhibits protein translation of EPHA2 mRNA by directly binding to its 3'UTR region and promotes the proliferation and lymph node metastasis in esophageal squamous cell carcinoma [24]. Uc.57 binds to BCL11A mRNA, which reduced tamoxifen resistance by inhibiting PI3K/AKT and MAPK signaling pathways [25]. All these studies were conducted in cancer models, where T-UCRs consistently act as negative regulators of mRNA expression. Uc.478 and uc.479 affect the expression of genes in non-cancerous conditions. Their overexpression is associated with decreased levels of GRIA3 mRNA and protein in human models of Rett syndrome [26].

Another interesting regulatory mechanism exercised by T-UCRs was presented by Zhang et al. [27]. The authors revealed a new regulatory relationship between T-UCRs and their host genes. They suggested that uc.189 positively regulates the expression of the splicing factor SRp20, but did not provide any mechanistic evidence to explain how this occurs [27].

3.2 Interaction between T-UCRs and proteins

Regarding the mechanisms of action of some T-UCRs, there is no evidence that they affect the expression of mRNAs, but functional analyzes show the levels of the corresponding proteins are significantly decreased when T-UCRs are overexpressed or increased when T-UCRs are silenced. This suggests a T-UCR-protein interaction. This is the case with uc.38 and PBX1. Zhang et al. found that uc.38 interacts with the protein and subsequently affects the expression of Bcl-2 family members, ultimately inducing apoptosis in breast cancer cells [28]. Shi et al. also revealed that T-UCRs can directly interact with proteins. In their study, uc.51 physically interacts with NONO and promotes the proliferation and migration of breast cancer cells [29].

As a type of lncRNAs, T-UCRs may have the ability to specifically bind with one or more proteins or function as scaffolds for protein complexes involved in the regulation of gene expression. Similar capabilities of T-UCRs were reported by Zambalde et al. who identified 17 proteins that interact with uc.147 either directly or indirectly [30]. For some of these interactions, Zambalde et al. suggested a role for uc.147 as a regulator of protein functions. The authors hypothesized that uc.147 could serve as a scaffold for two protein complexes: one between RHOA and CIT, and another among six proteins (WDR18, TEX10, PES1, NOP53, DKC1, and ZC3HAV1) that participate in RNA processing and regulation. Additionally, the authors found correlations between the expression of uc.147 and neighboring genes, as well as downstream UCRs uc.148 and uc.149. Zambalde et al. do not exclude the possibility that uc.147 also uses another regulatory mechanism in breast cancer—acting as a decoy for miR-190 [30].

When discussing the mechanisms of action of T-UCRs in regulation of the gene expression, it is important to mention also a less well-known lncRNA-dependent epigenetic mechanism, called by Wang et al. an "epigenetic switch point" [31]. Sun et al. discovered that uc.323 can directly interact with EZH2, which is the primary component of the histone modification complex PRC2 responsible for the N-methylation of histones. This interaction can inhibit H3K27me3 at the promoter regions of CPT1b and regulate its transcription. Thus uc.323 ameliorated cardiac hypertrophy [32].

Studies by Panatta et al. and Mancini et al. suggest that T-UCRs mediate another epigenetic mechanism. According to these authors, uc.291 modulates chromatin remodeling activity by negatively regulating ACTL6A, the catalytic subunit of the SWI/SNF chromatin remodeling complex. This releases chromatin and allows for the transcription of epidermal differentiation complex genes. In cutaneous squamous cell carcinoma, downregulation of uc.291 reduces SWI/SNF complex activity, leading to a de-differentiation phenotype [33, 34].

3.3 Interaction between T-UCRs and microRNAs

The interaction between T-UCRs and microRNAs (miRs) is another way for T-UCRs to exert their regulatory functions. This interaction can modulate the activity of miRs. The effects of this can be seen through the degradation or inactivation of T-UCRs' targets, miRs, and primary miRs (pre-miR), or through the influence on the expression of mature miRs [9].

Table 1 presents examples of T-UCRs that act as competitive endogenous RNAs (ceRNAs) and prevent binding of miRs to their target mRNAs in T-UCRs/miR/mRNA axis.

T-UCRs *ceRNAs*	**miRs/ mRNA axis**	**Role of T-UCRs in pathology**	**Expression/cancer type/disease/injury**	**References**
uc.8+	miR-596/ MMP9	Promotes invasion and metastasis	Increased in early stage of disease/BIC	[35]
uc.8+ uc.201+	miR-596	In combination inhibit tumor suppressor miR-596		[36]
uc.63+	miR-130b/ MMP2	Involves in cell growth and cell migration	Increased/PC	[37]
uc.77	miR-4676-5p/ FBXW8	Inhibites the proliferation tumor cells by increasing FBXW8 and CDK4 ubiquitination, which lead to negatively regulation of CDK4 and a block of the G0/G1 transition	Decreased/CRC	[38]
uc.158+A	miR-193b	Drives tumor growth and migration, its inhibition increased apoptosis	Increased/HCC Increased/CCA	[39]
uc.173	miR-29b/ CLDN1	Enhances the intestinal epithelial barrier function	*Intestinal mucosa*	[40]
	pre-miR-195	Stimulates intestinal epithelial regeneration, enhances growth of the intestinal mucosa in mice		[41]
	miR-291a-3p	Inhibits lead-induced neuronal apoptosis	Decreased/lead-induced nerve injury	[42]
uc.230	miR-503/ CUGBP1	Promotes renewal and barrier function of intestinal epithelia, protects against cell death, enhances mucosal repair	Increased/UC	[43]
uc.333	miR-223/ FOXO1	Improves insulin resistance	decreased/NAFLD	[44]

T-UCRs *ceRNAs*	miRs/ mRNA axis	Role of T-UCRs in pathology	Expression/cancer type/disease/injury	References
uc.339	miR-339-3p miR-663b-3p miR-95-5p	Promotes tumor growth and migration by upregulation of Cyclin E2, a direct target of these miRs	Increased/NSCLC	[45]
	pri-miR-339/ SLC7A11	Leads to defects in the ferroptosis, which drives the metastasis	Increased/LC adenocarcinoma	[46]
uc.372	pri-miR-195/ pri-miR-4668/ AGO	Promotes lipogenesis and lipid accumulation in liver by regulation of expression of *FAS* and *ACC/SCD1* and *CD36* genes by Ago2.	Increased/NAFLD	[47]
uc.416+A	miR-153	Affects cancer growth and migration	Increased/RC	[48]

Source: BlC—bladder cancer; CCA—cholangiocarcinoma; CRC—colorectal cancer; HCC—hepatocellular carcinoma; LC—lung cancer; NAFLD—nonalcoholic fatty liver disease; NSCLC—non-small cell lung cancer; PC—prostate cancer; RC—renal cancer; UC—ulcerative colitis.

Table 1.
T-UCRs as a decoy for miRs and their role in different diseases.

Another aspect of the interaction between T-UCRs and miRs is connected to the influence of T-UCRs on the expression of mature miRs.

Liz et al. describe a possible mechanism by which T-UCRs regulate the maturation of miRs [49]. In their study, they found that uc.283+A interacts with the stem region of pri-miR-195, preventing its processing by the Drosha/DGCR8 microprocessor complex. Similarly, Zhang et al. found that another T-UCR, uc.339, prevented the Drosha/DGCR8 splicing complex from binding to pri-miR-339, inhibiting the production of mature miR-339 [46]. Uc.372 regulates the cleavage of both pri-miR-195 and pri-miR-4668 from pri-to-pre-miR by sponging them, suppressing their maturation [47]. In contrast, a new study shows that uc.160+ enhances the processing of the pri-miR-376 cluster and A-to-I editing during miR-376 maturation [50].

Is the interaction between T-UCRs and miRs direct, or is there a mediator involved? This is another question that the study by Terreri et al. addresses [36]. The authors found that the interaction between T-UCRs and miRs can be mediated by the polycomb protein Yin Yang 1 (YY1). Specifically, YY1 mediates the binding between uc.8+ and miR-596 in bladder cancer. The authors suggest that the binding of YY1 to T-UCR may alter its conformation, inhibiting the binding of miR-596 [36]. This finding reveals a new player in the interaction between T-UCRs and miRs, another factor in gene expression regulation.

In their study, Terreri et al. shed another new light on the relationship between T-UCRs and miRs. They found that silencing uc.8+ increased the expression of miR-596, making it available to bind to other T-UCRs such as uc.283+, which in turn reduced its level. Terreri et al. believed that the interaction between T-UCRs and miRs occurs in networks that modulate the availability of these non-coding RNAs. The authors referred to this as an example of inter epigenetic regulation [36].

A recent study by Corrà et al. provided further evidence of this network interaction between T-UCRs and miRs in breast cancer [51]. The authors found that silencing uc.183 and uc.96 increased the levels of pre-miR-221 and miR-221, but also that miR-221 itself reduced the expression of uc.183, uc.110, and uc.84. This raises the question of which of the two has a regulatory function, T-UCRs or miR? This question is further complicated by the fact that the study found that treatment with anti-cancer drugs affects the expression of T-UCRs, but not the expression of miR-221.

Although the number of studies investigating the mechanisms of action of T-UCRs is increasing, their primary functions remain unclear. It is important to determine whether T-UCRs exhibit most of the functions described for other classes of lncRNAs or if new, specific functions will be discovered. A key feature that distinguishes T-UCRs from other lncRNAs is their bi-directional interaction with miRs—both regulating and being regulated by them.

There are also other studies that have reported on the role of T-UCRs in a number of pathological processes, but the underlying regulatory mechanistic interactions remain unclear and warrant further investigation [52–55].

Despite attempts to systematically present and summarize the mechanisms of action of T-UCRs, it cannot be categorically stated that there is a manifestation of a single mechanism during functional activity. Different manifestations of action from one T-UCR in different diseases cannot be ruled out, similar to other non-coding RNAs, although the trend in this regard with T-UCRs is significantly weaker. An idea of the diversity of mechanisms of action, their effects and the behavior of one T-UCR in different types of cancer can be given by uc.338, which is one of the most studied T-UCRs.

It is believed that uc.338 stimulates cell growth and promotes the transition from the G1 to S phase of the cell cycle, leading to increased cell migration and invasion in hepatocellular carcinoma, lung, colorectal, and cervical cancers. Uc.338 stimulates cell growth through the PI3K/AKT pathway, potentially by down-regulating p21 and up-regulating cyclin D1 [56], or in association with BMI1, by modulating the transcription of CDKN1A and partially repressing p21 [57]. Around 400 potential target genes with uc.338-binding sites were identified, many of which are involved in cell proliferation [58]. Moreover, PAI-RBP1 (plasminogen activator inhibitor-1 RNA-binding protein) was identified as a potential RNA-binding partner of uc.338 [58]. Additionally, as mentioned above, silencing uc.338 inhibits cell migration and invasion by directly regulating TIMP1 [19, 20], and after knocking down uc.338, p53 activity increased by almost 30% [58].

4. Mechanisms of regulation of T-UCRs expression

The expression of T-UCRs is regulated through several mechanisms, including miRs, CpG island methylation, and histone modifications. These mechanisms are all forms of epigenetic influence. In some cases, these mechanisms work together. For example, the expression of uc.160+ is simultaneously affected by miR-155 and the methylation of the host gene promoter [59, 60].

4.1 Role of microRNAs

It is crucial to further investigate the regulatory role of T-UCRs through miRs. Moreover, there is evidence indicating that miRs significantly influence the regulation of T-UCRs transcription, particularly in cancer.

MiRs as regulators of T-UCRs expression are presented mainly in cancer (**Table 2**). The majority of studies paid attention to the sequence complementarity between T-UCRs and miRs. This effect was first reported by Galin et al. who found that the

T-UCRs targets	miRs regulators	Cancer type	References
uc.160+	miR-155	CLL	[5]
		GC	[59]
	miR-24-1	CLL	[5]
uc.346+A	miR-155	CLL	[5]
uc.348+	miR-29b	CLL	[5]
uc.209+	miR-877-3p	Neuroblastoma	[61]
uc.271+	miR-383		
uc.312+	miR-877-3p		
	miR-548d-5p		
uc.330+	miR-548d-5p		
uc.371+	miR-877-3p		
uc.411+	miR-33b-5p		
uc.421+	miR-877-3p		
uc.435+	miR-939		
uc.452+	miR-383		
uc.416+A	miR-153	GC	[23]
uc.283	miR-596	BIC	[36]

Source: BIC—bladder cancer; CLL—chronic lymphocytic leukemia; GC—gastric cancer.

Table 2.
T-UCRs regulated through miRs in different type of cancer.

expression levels of two T-UCRs, uc.160+ and uc.346+A, could be directly regulated in vivo by a single miRNA, miR-155, in chronic lymphocytic leukemia [5]. The role of miR-155 as a regulator of uc.160+ has also been confirmed in gastric cancer [59]. In neuroblastoma, nine T-UCRs have been suggested to be regulated by five miRNAs [61]. However, the role of miRNAs as regulators of T-UCR expression remains uncertain. This is evidenced by research conducted by the group led by Wataru Yasui, who found in 2016 that miR-153 directly suppresses uc.416+A in a model of gastric cancer [23], but in 2018 claimed that uc.416+A modulates epithelial-to-mesenchymal transition through the regulation of miR-153 in renal cell carcinoma [48]. The effect of miR-596 on uc.283 expression discussed above [36] also indicates that there is no clear, unambiguous mechanism linking miRNAs and T-UCRs.

4.2 Role of CpG island methylation

Another way in which T-UCRs expression is regulated is through the hypermethylation of CpG islands in their promoter regions. This effect, which results in the downregulation and inactivation of three T-UCRs (uc.160+, uc.283+A, and uc.346+), was found by Lujambio et al. [62] and conformed for uc.160+ expression by Honma et al. [60]. They also revealed that silencing of uc.160+ led to the activation of MAPK signaling via repression of PTEN expression. Hudson et al. in turn confirmed the silencing of uc.283+A due to CpG islands hypermethylation [62]. Moreover, they with Lujambio et al. reported that treatment with DNA hypomethylating agent

5-aza-2-deoxycytidine (5-Aza-dC) upregulated the T-UCRs expression in human colon cancer and prostate cancer cells [62]. Goto et al. provided further evidence for this mechanism by discovering that in gastric and prostate cancer, uc.158+A is downregulated due to DNA methylation in the promoter region upstream of its transcription gene [23]. Hudson et al. discovered that five additional T-UCRs (uc.308+A, uc.434+A, uc.241+A, uc.285+, and uc.85+) were epigenetically downregulated, and their expression could be restored by treatment with a combination of 5-Aza-dC and histone deacetylase inhibitor trichostatin A (TSA) [63]. It seems that this mechanism, of influencing the expression of T-UCRs, leads to the silencing of those T-UCRs that have tumor-suppressive functions.

4.3 Role of histone mark

According to Mestdagh et al. both intergenic and intragenic T-UCRs are marked by trimethylation of lysine 4 on histone H3 (H3K4me3), a marker for transcriptional initiation [6]. The authors compared the distribution of active H3K4me3 marks between T-UCRs, miRNAs, and protein-coding genes in order to understand the structure and function of T-UCR transcriptional units. They found that the distribution of active H3K4me3 marks in T-UCRs is different from that of protein-coding genes, but it shows a remarkable resemblance to that of miRs. As a result, the authors hypothesized that there may be commonalities in the transcriptional organization between these two classes of ncRNAs—T-UCRs and miRs, and differences between T-UCRs and protein-coding genes [6].

4.4 Role of other factors

The studies by Vannini et al. and Guo et al. present some established influences on the expression of T-UCRs, but the underlying mechanisms remain unknown [45, 47]. Vannini et al., observed that the tumor-suppressor TP53 directly and negatively regulates uc.339. They proposed that the deregulation of uc.339 and its manifestation as an oncogene may be caused by mutations in the TP53 gene, which frequently occurs in non-small cell lung cancer. In their study, Guo et al., discovered that the upstream transcription factor 1 (USF1) regulates the expression of uc.372 in non-alcoholic fatty liver disease. They also noted that INSM2, which shares a promoter with uc.372, is involved in this regulation. The authors found that the knockdown of INSM2 decreases the level of uc.372 and suppresses USF1's regulation of uc.372. The regulation of T-UCRs expression by transcription factors has also been described in more recent studies on uc.147 and uc.43+ [64, 65].

Another factor affecting T-UCRs transcription is hypoxia. Hypoxia can induce T-UCRs such as uc.475+, uc.63+/+A, uc.73+A, uc.106+, and uc.134+A (also known as hypoxia-induced noncoding ultraconserved transcripts, HINCUTs) [66]. While HIF-1α-binding sites have been discovered upstream promoter regions of uc.475+ and uc.63+, the mechanism by which other HINCUTs are induced by hypoxia remains unknown [67].

5. The influence of genetic variants in UCRs

Despite their strong conservatism, UCRs are not resistant to variations. In 2010, Wojcik et al., claimed that UCRs are sites of very low natural variation in the general population [68]. De Grassi et al. argued that UCRs accumulate fewer mutations overall, including fewer nearly neutral mutations [69]. Depletion of prevalent neutral SNPs in UCRs was also confirmed by Silla et al. [70]. Halligan et al. argued that UCRs receive more strongly selected deleterious mutations than protein-coding genes or nearby regulatory elements [71]. Habic et al. unlike Wojcik et al., and De Grassi et al., found that the polymorphism density within UCRs is higher than the genomic average, but in support of Halligan et al. believe that SNPs in UCRs may be less stable or rarer, with deleterious consequences in populations than those outside of UCRs [72]. However, only a small portion of the established SNPs in UCRs have a proven association with human diseases. It is interesting to note that many mutations within UCRs are not fixed [16]. Habic et al. propose that not all of these regions are subject to the same level of purifying selection [72]. Silla et al. suggest that UCRs are not under selective pressure as a continuous stretch of DNA, but rather experience differential evolutionary pressure at the single nucleotide level [70]. Therefore, the question arises, can the hypothesis that UCRs are subject to stronger negative selection [73] be considered as a general principle for UCRs?

In their paper, Habic et al. [72] summarize the studies in the literature about the associations between the presence of SNPs in UCRs and the appearance of different disorders. They found 37 polymorphism/phenotype associations covering diseases such as: different types of muscular dystrophies; adolescent idiopathic scoliosis; amyotrophic lateral sclerosis; renal diseases; as well as eye-related diseases; and cancer [72]. Few studies examining the association between SNPs in UCRs and disease mention that the analyzed polymorphisms are located within conserved genomic regions. This makes it challenging to organize and categorize them systematically.

Studying SNPs in UCRs presents an opportunity to easily detect disease biomarkers. Several studies have established an association between individual SNPs in UCRs and disease risk, especially cancer risk [74–76]. Additionally, SNPs within UCRs may serve as valuable prognostic and predictive biomarkers, as demonstrated in studies by Lin et al. and Bao et al. [76, 77]. Rs8004379 SNP in uc.368+ is significantly associated with biochemical recurrence in localized prostate cancer as well as a decreased risk for prostate cancer-specific mortality in patients with advanced disease. In their analysis of the functional consequences of this SNP, Bao et al. made an interesting conclusion [77]. Given the localization of rs8004379 within an intron of the tumor suppressor *NPAS3* gene, they suggested that the rs8004379 C allele may affect the expression of *NPAS3* and be associated with a decreased risk of recurrence and mortality in prostate cancer patients after treatment. Rs7849 in uc.298 in a study by Lin et al. was found to be associated with an increased risk of disease recurrence in stage II patients [76]. Also, rs10211390 in uc.54 is associated with an increased risk of recurrence after 5-fluorouracil-based chemotherapy in patients with stage II and III colorectal cancer. The authors believed that individual outcomes following chemotherapy can be predicted based on the rs7849 and rs6590611 genotypes [76].

6. Clinical significance of T-UCRs in cancer and other diseases

In recent years, the number of studies focused on T-UCRs has notably grown. A significant part of them includes functional analyses to confirm the impact of T-UCRs in disease pathogenesis. However, these studies often do not provide details about the specific mechanisms of action and direct interactions of these T-UCRs [5, 39, 67, 78–81]. Another large group of studies is descriptive, providing information about the association of T-UCRs levels with clinicopathological features of patients or reporting changes in T-UCRs expression in response to therapy. The results of this second group of studies provide an opportunity to select T-UCRs as potential diagnostic, prognostic, and predictive biomarkers. A large portion of these studies is summarized in **Table 3**, where information about the type of disease, the expression levels of T-UCRs, and their association with the clinical characteristics of studied patients is presented.

T-UCRs/ expression	Cancer type/ disease	Association with clinicopathological characteristics	Possible biomarker	Reference
uc.8+/ downregulated	BlC	Pathological grade pathological stage worst overall survival tumors at advanced stages—T-UCR prevalent in cytoplasma	Prognostic	[35, 36]
uc.38/ downregulated	BC	Larger tumors tumors of more advanced stages	—	[28]
		T3 stage stage IV expression—ER, PR and HER2 receptors	—	[82]
uc.48+/ upregulated	TB	Distinguish between untreated and cured patients	Diagnostic	[53]
uc.51/ upregulated	BC	Larger tumor size metastasis	—	[29]
uc.57/ downregulated	BC	Tamoxifen resistance	Predictive	[25]
uc.63/ upregulated	Luminal A BC	Reduced disease-free survival	Prognostic	[83]
	BC	M1 trait/TNM classification expression of the HER2 receptor	—	[82]
	MIBC	Knockdown of uc.63+ increased sensitivity to cisplatin chemotherapy	Predictive	[84]
	PC	Castration resistance, docetaxel resistance, worse prognosis	Serum predictive, prognostic	[37]
	LC	Tumor stage, poor prognosis	Prognostic	[85]
	GC	Advanced pT3/4 stages, pN1/2/3 stages, stages III/IV	Prognostic	[67]
uc.70/ upregulated	CLL	Inferior clinical outcome, downregulated by CpG-ODN treatment and predict longer time to treatment	Prognostic	[86]
uc.73/ downregulated	CRC	Worse overall survival	Prognostic	[79]
uc.84/ upregulated	BC	Less aggressive tumors worse overall survival	Prognostic	[87]
uc.98/ upregulated	AS/ACS	Atherosclerotic plaque instability	Diagnostic, prognostic	[54]
uc.134/ downregulated	HCC	Worse overall survival	Prognostic	[88]
uc.147/ upregulated	BC	Worse overall survival worse progression-free survival in HER2+ patients	Prognostic	[87]

T-UCRs/ expression	Cancer type/ disease	Association with clinicopathological characteristics	Possible biomarker	Reference
uc.160 methylated/ upregulated	CRC	Lymph node infiltration, lesion size	—	[89]
		Stages III/IV better overall survival	Prognostic	[90]
	Lower-grade gliomas	Better overall survival	Prognostic	[50]
uc.189/ upregulated	CSCC EAC	Poor prognosis	Prognostic	[91]
	ESCC	Invasion, advanced clinical stage, lymph node metastasis, poor prognosis	Prognostic	[92]
uc.261/ upregulated	CD	Positively correlated with HBI	—	[80]
uc.268/ upregulated	Luminal A BC	Poor survival rate	Prognostic	[87]
uc.283 methylated/ upregulated	CRC	Better overall survival	Prognostic	[90]
uc.306/ downregulated	HBV-related HCC	Shorter overall survival	Prognostic	[93]
uc.338/ upregulated	CRC	Distal colon location	—	[79]
		Lymph node metastasis	—	[19]
		Larger tumor size, lymph node metastasis, poorer prognosis	Prognostic	[56]
	NSCLC	TNM stage, lymph node metastasis, distant metastasis, shorter overall survival, shorter disease-free survival	Prognostic	[94]
	CC	With lymph node metastasis	—	[20]
uc.454/ downregulated	NSCLC	Lymph node metastasis	—	[22]

Source: ACS—acute coronary syndrome; ALL—acute lymphoblastic leukemia; AS—atherosclerosis; BC—breast cancer; BlC—bladder cancer; CC—cervical cancer; CD—Crohn's disease; CLL—chronic lymphocytic leukemia; CpG ODN—oligodeoxynucleotides that contain unmethylated CG dinucleotides; CRC—colorectal cancer; CSCC—cervical squamous cell carcinomas; EAC—endometrial adenocarcinomas; EC—esophagus cancer; ESCC—oesophagal squamous cell carcinomas; GC—gastric cancer; HBI—Harvey-Bradshaw Index (to evaluate the activity and severity of CD and to monitor the effect of treatment); HBV—hepatitis B-related HCC; HCC—hepatocellular carcinoma; LC—lung cancer; MIBC—muscle-invasive bladder cancer; NSCLC—non-small cell lung cancer; TB—tuberculosis; PC—prostate cancer.

Table 3.
T-UCRs with clinical, predictive and prognostic significance.

Examining the functions of T-UCRs individually and searching for a single target for one T-UCR is an approach that is increasingly being replaced by more complex ones, such as constructing three-component regulatory networks—T-UCRs-miRNAs-mRNAs. Simultaneously tracking multiple T-UCR interactions within a single disease allows for the evaluation of the potential diagnostic and prognostic values of T-UCRs, as demonstrated in the study by Khalafiyan et al., where five T-UCRs including uc.8, uc.169, uc.342, uc.360, and uc.417 were found to be correlated with the overall survival of gastric cancer patients [95]. On the other hand, three-component regulatory networks provide an opportunity for a comprehensive understanding of the underlying gene expression regulation processes involved in pathogenesis.

Calin et al. first reported that dysregulation of several T-UCRs is associated with disease [5]. Later, Scaruffi et al. presented two T-UCR panels for high-risk prediction in neuroblastoma patients [61]. The first panel included 28 T-UCRs significantly associated with prognosis, while the second panel of 15 upregulated T-UCRs discriminated short- from long-survivors with high sensitivity and specificity. In 2014, Fassan et al. presented two panels of T-UCRs for biological profiling in the transformation of Barrett's mucosa [96]. One is for distinguishing Barrett's esophagus from the squamous epithelium and another is for evaluating progression from normal to adenocarcinoma. Zambalde et al. presented two panels of T-UCRs as diagnostic biomarkers for breast cancer and a third panel that distinguished with high sensitivity and specificity luminal A from triple-negative patients [87]. Finally, we come to studies such as that by Ciaramella et al. in which the authors propose a new biomarker panel of T-UCRs using machine learning techniques [97]. This signature panel of 13 T-UCRs is able to efficiently discriminate between normal and bladder cancer patient samples or patients with different survival extents, identifying sub-types of bladder cancer patients with different prognoses regardless of cancer grade. The implementation of Artificial Intelligence methodologies in medical applications represents a significant challenge in our time.

7. Future perspectives

As the number of studies on T-UCRs continues to grow, it is expected that more specific information will be uncovered about their structure, functions, and potential therapeutic applications. Understanding how T-UCRs operate in normal cells will provide insight into why and how genomic ultraconservation is maintained. A key challenge is to identify the role of T-UCRs in maintaining tissue homeostasis. By clarifying their role and mechanism of action in normal cells, it may be possible to understand the consequences of their dysregulation in pathological conditions. Increasingly, research is focusing on T-UCRs as potential molecular biomarkers or as new drug targets, particularly in cancer treatment. As biomarkers, T-UCRs are promising for both diagnosis and assessing prognosis because it is known that for some T-UCRs, the levels of expression and their localization within the cell vary at different stages of cancer. T-UCRs also hold potential as a basis for therapeutic strategies or drug combinations. The therapies are aimed at destroying cancer cells or reversing the tumor phenotype, as well as influencing mechanisms related to resistance.

Research on T-UCRs requires a complex approach due to the diverse interactions they are involved in and the intricate mechanisms regulating their expression.

8. Conclusions

In nearly 20 years since the first description of UCRs by Bejerano et al., scientific interest and research on them and T-UCRs have grown significantly. The interest directed towards T-UCRs is justifiably great, as their physiological significance in both normal and pathological conditions has been demonstrated. However, our understanding of their functions and the mechanisms through which they are realized still sheds little light on "dark matter". T-UCRs, as a type of lncRNA, exhibit more characteristic features than common functions with other ncRNAs. Their interactions with miRs or proteins reveal the richness of the genetic regulatory machinery, but also provide an opportunity for its precise influence through the development of new RNA-based therapies.

Acknowledgements

This work was funded by the Bulgarian National Science Fund (Grand number КП-06-H23/6, 18.12.2018).

Conflict of interest

The author declares no conflict of interest.

Abbreviations

CIT	citron rho-interacting serine/threonine kinase
CPT1b	carnitine palmitoyl transferase 1b
DKC1	dyskerin pseudouridine synthase 1
EPHA2	ephrin type-A receptor 2
EZH2	enhancer of zeste homolog 2
HOXB5	Homeobox B5
HSPA12B	heat shock protein family A member 12B
IGFBP1	insulin-like growth factor-binding protein 1
INSM2	insulinoma-associated 2 gene
MMP9	matrix metalloproteinase-9
NONO	non-POU domain-containing octamer-binding protein
NOP53	glioma tumor suppressor candidate region gene 2 protein
PBX1	pre-B-cell leukemia homeobox 1
PES1	Pescadillo Ribosomal Biogenesis factor 1
POM121	transmembrane nucleoporin pseudogene
PRC2	polycomb repressor complex 2
RHOA	Ras homolog family member A
TEX10	testis expressed 10
TIMP1	tissue inhibitor of metalloproteinase 1
WDR18	WD Repeat Domain 18
ZC3HAV1	Zinc finger antiviral 1

Author details

Maria Radanova
Medical University of Varna, Varna, Bulgaria

*Address all correspondence to: maria.radanova@mu-varna.bg

References

[1] Braicu C, Zimta AA, Harangus A, Iurca I, Irimie A, Coza O, et al. The function of non-coding RNAs in lung cancer tumorigenesis. Cancers (Basel). 2019;**11**(5):605. DOI: 10.3390/cancers11050605

[2] Bresesti C, Vezzoli V, Cangiano B, Bonomi M. Long non-coding RNAs: Role in testicular cancers. Frontiers in Oncology. 2021;**11**:605606. DOI: 10.3389/fonc.2021.605606

[3] Bejerano G, Pheasant M, Makunin I, Stephen S, Kent WJ, Mattick JS, et al. Ultraconserved elements in the human genome. Science. 2004;**304**:1321-1325. DOI: 10.1126/science.1098119

[4] Gao SS, Zhang ZK, Wang XB, Ma Y, Yin GQ, Guo XB. Role of transcribed ultraconserved regions in gastric cancer and therapeutic perspectives. World Journal of Gastroenterology. 2022;**28**(25):2900-2909. DOI: 10.3748/wjg.v28.i25.2900

[5] Calin GA, Liu CG, Ferracin M, Hyslop T, Spizzo R, Sevignani C, et al. Ultraconserved regions encoding ncRNAs are altered in human leukemias and carcinomas. Cancer Cell. 2007;**12**(3):215-229. DOI: 10.1016/j.ccr.2007.07.027

[6] Mestdagh P, Fredlund E, Pattyn F, Rihani A, Van Maerken T, Vermeulen J, et al. An integrative genomics screen uncovers ncRNA T-UCR functions in neuroblastoma tumours. Oncogene. 2010;**29**(24):3583-3592. DOI: 10.1038/onc.2010.106

[7] Terracciano D, Terreri S, de Nigris F, Costa V, Calin GA, Cimmino A. The role of a new class of long noncoding RNAs transcribed from ultraconserved regions in cancer. Biochimica Et Biophysica Acta. Reviews on Cancer. 2017;**1868**(2):449-455. DOI: 10.1016/j.bbcan.2017.09.001

[8] Fabris L, Calin GA. Understanding the genomic Ultraconservations: T-UCRs and cancer. International Review of Cell and Molecular Biology. 2017;**333**:159-172. DOI: 10.1016/bs.ircmb.2017.04.004

[9] Chava S, Challagundla KB. The transcribed-ultra conserved Regions: Novel non-coding RNA players in neuroblastoma progression. Noncoding RNA. 2019;**5**(2):39. DOI: 10.3390/ncrna5020039

[10] Pereira Zambalde E, Mathias C, Rodrigues AC, de Souza Fonseca Ribeiro EM, Fiori Gradia D, Calin GA, et al. Highlighting transcribed ultraconserved regions in human diseases. Wiley Interdisciplinary Reviews RNA. 2020;**11**(2):e1567. DOI: 10.1002/wrna.1567

[11] Wang Y, Zhou D, Feng Y, Chen G, Li N. T-UCRs with digestive and respiratory diseases. Bioorganic & Medicinal Chemistry Letters. 2020;**30**(16):127306. DOI: 10.1016/j.bmcl.2020.127306

[12] Bozgeyik I. The dark matter of the human genome and its role in human cancers. Gene. 2022;**811**:146084. DOI: 10.1016/j.gene.2021.146084

[13] Gibert MK Jr, Sarkar A, Chagari B, Roig-Laboy C, Saha S, Bednarek S, et al. Transcribed ultraconserved regions in cancer. Cells. 2022;**11**(10):1684. DOI: 10.3390/cells11101684

[14] Wang T, Li F, Lu Z. Ultra-conserved RNA: A novel biological tool with diagnostic and therapeutic potential.

Discover Oncology. 2023;**14**(1):41. DOI: 10.1007/s12672-023-00650-1

[15] Oliveira JC. Transcribed Ultraconserved Regions: New regulators in cancer signaling and potential biomarkers. Genetics and Molecular Biology. 2023;**46**(1 Suppl. 2):e20220125. DOI:10.1590/1678-4685-GMB-2022-0125

[16] Fedorova L, Mulyar OA, Lim J, Fedorov A. Nucleotide composition of ultra-conserved elements shows excess of GpC and depletion of GG and CC dinucleotides. Genes (Basel). 2022;**13**(11):2053. DOI: 10.3390/genes13112053

[17] Irimie AI, Zimta AA, Ciocan C, Mehterov N, Dudea D, Braicu C, et al. The unforeseen non-coding RNAs in head and neck cancer. Genes (Basel). 2018;**9**(3):134. DOI: 10.3390/genes9030134

[18] Li Q, Li X, Wang C. Uc.206 regulates cell proliferation and apoptosis by targeting P53 in cervical cancer cells. Neoplasma. 2016;**63**(3):411-418. DOI: 10.4149/310_151017N538

[19] Wang C, Wang Z, Zhou J, Liu S, Wu C, Huang C, et al. TUC.338 promotes invasion and metastasis in colorectal cancer. International Journal of Cancer. 2017;**140**(6):1457-1464. DOI: 10.1002/ijc.30542

[20] Li Q, Shen F, Wang C. TUC338 promotes cell migration and invasion by targeting TIMP1 in cervical cancer. Oncology Letters. 2017;**13**(6):4526-4532. DOI: 10.3892/ol.2017.5971

[21] Zhou J, Wang C, Gong W, Wu Y, Xue H, Jiang Z, et al. Uc.454 inhibited growth by targeting heat shock protein family a member 12B in non-small-cell lung cancer. Molecular Therapy - Nucleic Acids. 2018;**12**:174-183. DOI: 10.1016/j.omtn.2018.05.004

[22] Zhou J, Wang C, Huang C, Ding Z, Shi M. TUCR.454 inhabits metastasis in lung cancer cells. Int. International Journal of Clinical and Experimental Pathology. 2018;**11**(3):1289-1296

[23] Goto K, Ishikawa S, Honma R, Tanimoto K, Sakamoto N, Sentani K, et al. The transcribed-ultraconserved regions in prostate and gastric cancer: DNA hypermethylation and microRNA-associated regulation. Oncogene. 2016;**35**(27):3598-3606. DOI: 10.1038/onc.2015.445

[24] Ding Z, Yan Y, Guo YL, Wang C. Esophageal carcinoma cell-excreted exosomal uc.189 promotes lymphatic metastasis. Aging (Albany NY). 2021;**13**(10):13846-13858. DOI: 10.18632/aging.202979

[25] Zhang CH, Wang J, Zhang LX, Lu YH, Ji TH, Xu L, et al. Shikonin reduces tamoxifen resistance through long non-coding RNA uc.57. Oncotarget. 2017;**8**(51):88658-88669. DOI: 10.18632/oncotarget.20809

[26] Siqueira E, Obiols-Guardia A, Jorge-Torres OC, Oliveira-Mateos C, Soler M, Ramesh-Kumar D, et al. Analysis of the circRNA and T-UCR populations identifies convergent pathways in mouse and human models of Rett syndrome. Molecular Therapy - Nucleic Acids. 2021;**27**:621-644. DOI: 10.1016/j.omtn.2021.12.030

[27] Zhang M, Zhou J, Jiao L, Xu L, Hou L, Yin B, et al. Long non-coding RNA T-uc.189 modulates neural progenitor cell fate by regulating Srsf3 during mouse cerebral cortex development. Frontiers in Neuroscience. 2021;**15**:709684. DOI: 10.3389/fnins.2021.709684

[28] Zhang LX, Xu L, Zhang CH, Lu YH, Ji TH, Ling LJ. Uc.38 induces breast cancer cell apoptosis via PBX1. American Journal of Cancer Research. 2017;**7**(12):2438-2451

[29] Shi X, Huang X, Chen R, Li Y, Xu Y, Zhang W, et al. The transcribed ultraconserved element uc.51 promotes the proliferation and metastasis of breast cancer by stabilizing NONO. Clinical & Experimental Metastasis. 2021;**38**(6):551-571. DOI: 10.1007/s10585-021-10128-5

[30] Zambalde EP, Bayraktar R, Schultz Jucoski T, Ivan C, Rodrigues AC, Mathias C, et al. A novel lncRNA derived from an ultraconserved region: Lnc-uc.147, a potential biomarker in luminal a breast cancer. RNA Biology. 2021;**18**(Suppl. 1):416-429. DOI: 10.1080/15476286.2021.1952757

[31] Wang Z, Zhang XJ, Ji YX, Zhang P, Deng KQ, Gong J, et al. The long noncoding RNA Chaer defines an epigenetic checkpoint in cardiac hypertrophy. Nature Medicine. 2016;**22**(10):1131-1139. DOI: 10.1038/nm.4179

[32] Sun Y, Fan W, Xue R, Dong B, Liang Z, Chen C, et al. Transcribed ultraconserved regions, Uc.323, ameliorates cardiac hypertrophy by regulating the transcription of CPT1b (carnitine palmitoyl transferase 1b). Hypertension. 2020;**75**(1):79-90. DOI: 10.1161/HYPERTENSIONAHA.119.13173

[33] Panatta E, Lena AM, Mancini M, Smirnov A, Marini A, Delli Ponti R, et al. Long non-coding RNA uc.291 controls epithelial differentiation by interfering with the ACTL6A/BAF complex. EMBO Reports. 2020;**21**(3):e46734. DOI: 10.15252/embr.201846734

[34] Mancini M, Cappello A, Pecorari R, Lena AM, Montanaro M, Fania L, et al. Involvement of transcribed lncRNA uc.291 and SWI/SNF complex in cutaneous squamous cell carcinoma. Discov. Oncologia. 2021;**12**(1):14. DOI: 10.1007/s12672-021-00409-6

[35] Olivieri M, Ferro M, Terreri S, Durso M, Romanelli A, Avitabile C, et al. Long non-coding RNA containing ultraconserved genomic region 8 promotes bladder cancer tumorigenesis. Oncotarget. 2016;**7**(15):20636-20654. DOI: 10.18632/oncotarget.7833

[36] Terreri S, Durso M, Colonna V, Romanelli A, Terracciano D, Ferro M, et al. New cross-talk layer between Ultraconserved non-coding RNAs, MicroRNAs and Polycomb protein YY1 in bladder cancer. Genes (Basel). 2016;**7**(12):127. DOI: 10.3390/genes7120127

[37] Sekino Y, Sakamoto N, Goto K, Honma R, Shigematsu Y, Sentani K, et al. Transcribed ultraconserved region uc.63+ promotes resistance to docetaxel through regulation of androgen receptor signaling in prostate cancer. Oncotarget. 2017;**8**(55):94259-94270. DOI: 10.18632/oncotarget.21688

[38] Zheng Z, Hong D, Zhang X, Chang Y, Sun N, Lin Z, et al. uc.77-downregulation promotes colorectal cancer cell proliferation by inhibiting FBXW8-mediated CDK4 protein degradation. Front. Oncologia. 2021;**11**:673223. DOI: 10.3389/fonc.2021.673223

[39] Carotenuto P, Fassan M, Pandolfo R, Lampis A, Vicentini C, Cascione L, et al. Wnt signalling modulates transcribed-ultraconserved regions in hepatobiliary cancers. Gut. 2017;**66**(7):1268-1277. DOI: 10.1136/gutjnl-2016-312278

[40] Wang JY, Cui YH, Xiao L, Chung HK, Zhang Y, Rao JN, et al.

Regulation of intestinal epithelial barrier function by long noncoding RNA uc.173 through interaction with MicroRNA 29b. Molecular and Cellular Biology. 2018;**38**(13):e00010-e00018. DOI: 10.1128/MCB.00010-18

[41] Xiao L, Wu J, Wang JY, Chung HK, Kalakonda S, Rao JN, et al. Long noncoding RNA uc.173 promotes renewal of the intestinal mucosa by inducing degradation of MicroRNA 195. Gastroenterology. 2018;**154**(3):599-611. DOI: 10.1053/j.gastro.2017.10.009

[42] Nan A, Zhou X, Chen L, Liu M, Zhang N, Zhang L, et al. A transcribed ultraconserved noncoding RNA, Uc.173, is a key molecule for the inhibition of lead-induced neuronal apoptosis. Oncotarget. 2016;**7**(1):112-124. DOI: 10.18632/oncotarget.6590

[43] Yu TX, Kalakonda S, Liu X, Han N, Chung HK, Xiao L, et al. Long noncoding RNA uc.230/CUG-binding protein 1 axis sustains intestinal epithelial homeostasis and response to tissue injury. JCI Insight. 2022;**7**(19):e156612. DOI: 10.1172/jci.insight.156612

[44] Zhang Y, Sun J, Yao H, Lin Y, Wei J, Hu G, et al. Ultraconserved element uc.333 increases insulin sensitivity by binding to miR-223. Aging (Albany NY). 2020;**12**(8):6667-6679. DOI: 10.18632/aging.103020

[45] Vannini I, Wise PM, Challagundla KB, Plousiou M, Raffini M, Bandini E, et al. Transcribed ultraconserved region 339 promotes carcinogenesis by modulating tumor suppressor microRNAs. Nature Communications. 2017;**8**(1):1801. DOI: 10.1038/s41467-017-01562-9

[46] Zhang N, Huang J, Xu M, Wang Y. LncRNA T-UCR uc.339/miR-339/SLC7A11 Axis regulates the metastasis of ferroptosis-induced lung adenocarcinoma. Journal of Cancer. 2022;**13**(6):1945-1957. DOI: 10.7150/jca.65017

[47] Guo J, Fang W, Sun L, Lu Y, Dou L, Huang X, et al. Ultraconserved element uc.372 drives hepatic lipid accumulation by suppressing miR-195/miR4668 maturation. Nature Communications. 2018;**9**(1):612. DOI: 10.1038/s41467-018-03072-8

[48] Sekino Y, Sakamoto N, Goto K, Honma R, Shigematsu Y, Quoc TP, et al. Uc.416 + a promotes epithelial-to-mesenchymal transition through miR-153 in renal cell carcinoma. BMC Cancer. 2018;**18**(1):952. DOI: 10.1186/s12885-018-4863-y

[49] Liz J, Portela A, Soler M, Gómez A, Ling H, Michlewski G, et al. Regulation of pri-miRNA processing by a long noncoding RNA transcribed from an ultraconserved region. Molecular Cell. 2014;**55**(1):138-147. DOI: 10.1016/j.molcel.2014.05.005

[50] Soler M, Davalos V, Sánchez-Castillo A, Mora-Martinez C, Setién F, Siqueira E, et al. The transcribed ultraconserved region uc.160+ enhances processing and A-to-I editing of the miR-376 cluster: Hypermethylation improves glioma prognosis. Molecular Oncology. 2022;**16**(3):648-664. DOI: 10.1002/1878-0261.13121

[51] Corrà F, Crudele F, Baldassari F, Bianchi N, Galasso M, Minotti L, et al. UC.110, and UC.84 ultra-conserved RNAs are mutually exclusive with miR-221 and are engaged in the cell cycle circuitry in breast cancer cell lines. Genes (Basel). 2021;**12**(12):1978. DOI: 10.3390/genes12121978

[52] Zhang Q, Cheng Z, Yu Z, Zhu C, Qian L. Role of lncRNA uc.457

in the differentiation and maturation of cardiomyocytes. Molecular Medicine Reports. 2019;**19**(6):4927-4934. DOI: 10.3892/mmr.2019.10132

[53] Li ZB, Han YS, Wei LL, Shi LY, Yi WJ, Chen J, et al. Screening and identification of plasma lncRNAs uc.48+ and NR_105053 as potential novel biomarkers for cured pulmonary tuberculosis. International Journal of Infectious Diseases. 2020;**92**:141-150. DOI: 10.1016/j.ijid.2020.01.005

[54] Fan Z, Zhang Y, Xiao D, Ma J, Liu H, Shen L, et al. Long noncoding RNA UC.98 stabilizes atherosclerotic plaques by promoting the proliferation and adhesive capacity in murine aortic endothelial cells. Acta Biochimica et Biophysica Sinica (Shanghai). 2020;**52**(2):141-149. DOI: 10.1093/abbs/gmz155

[55] Yu M, Guan Z, Li S, Wen X, Shi H, Qu G, et al. Gene expression profiling analysis reveals that the long non-coding RNA uc.412 is involved in mesangial cell proliferation. Molecular Medicine Reports. 2019;**20**(6):5297-5303. DOI: 10.3892/mmr.2019.10753

[56] Zhang Y, Wang S, Qian W, Ji D, Wang Q, Zhang Z, et al. Uc.338 targets p21 and cyclin D1 via PI3K/AKT pathway activation to promote cell proliferation in colorectal cancer. Oncology Reports. 2018;**40**(2):1119-1128. DOI: 10.3892/or.2018.6480

[57] Bo C, Li N, Li X, Liang X, An Y. Long noncoding RNA uc.338 promotes cell proliferation through association with BMI1 in hepatocellular carcinoma. Human Cell. 2016;**29**(4):141-147. DOI: 10.1007/s13577-016-0140-z

[58] Wen HJ, Walsh MP, Yan IK, Takahashi K, Fields A, Patel T. Functional modulation of gene expression by Ultraconserved long non-coding RNA TUC338 during growth of human hepatocellular carcinoma. iScience. 2018;**2**:210-220. DOI: 10.1016/j.isci.2018.03.004

[59] Pang L, Li Q, Zhang Y, Deng B, Wu F, Wang J, et al. Transcribed ultraconserved noncoding RNA uc.160 acts as a negative regulator in gastric cancer. American Journal of Translational Research. 2018;**10**(9):2822-2833

[60] Honma R, Goto K, Sakamoto N, Sekino Y, Sentani K, Oue N, et al. Expression and function of uc.160+, a transcribed ultraconserved region, in gastric cancer. Gastric Cancer. 2017;**20**(6):960-969. DOI: 10.1007/s10120-017-0714-9

[61] Scaruffi P, Stigliani S, Moretti S, Coco S, De Vecchi C, Valdora F, et al. Transcribed-ultra conserved region expression is associated with outcome in high-risk neuroblastoma. BMC Cancer. 2009;**9**:441. DOI: 10.1186/1471-2407-9-441

[62] Lujambio A, Portela A, Liz J, Melo SA, Rossi S, Spizzo R, et al. CpG island hypermethylation-associated silencing of non-coding RNAs transcribed from ultraconserved regions in human cancer. Oncogene. 2010;**29**(48):6390-6401. DOI: 10.1038/onc.2010.361

[63] Hudson RS, Yi M, Volfovsky N, Prueitt RL, Esposito D, Volinia S, et al. Transcription signatures encoded by ultraconserved genomic regions in human prostate cancer. Molecular Cancer. 2013;**12**:13. DOI: 10.1186/1476-4598-12-13

[64] Rodrigues AC, Zambalde EP, Bellan DL, Trindade EDS, Ribeiro EMSF, Calin G, et al. Lnc-uc.147 is associated with disease stage of liver, gastric,

and renal cancer. Biomolecules. 2023;**13**(2):265. DOI: 10.3390/biom13020265

[65] Ding L, Gong C, Zhao J, Liu X, Li T, Rao S, et al. Noncoding transcribed ultraconserved region (T-UCR) UC.48+ is a novel regulator of high-fat diet induced myocardial ischemia/reperfusion injury. Journal of Cellular Physiology. 2019;**234**(6):9849-9861. DOI: 10.1002/jcp.27674

[66] Ferdin J, Nishida N, Wu X, Nicoloso MS, Shah MY, Devlin C, et al. HINCUTs in cancer: Hypoxia-induced noncoding ultraconserved transcripts. Cell Death and Differentiation. 2013;**20**(12):1675-1687. DOI: 10.1038/cdd.2013.119

[67] Sakamoto N, Sekino Y, Fukada K, Pham QT, Honma R, Taniyama D, et al. Uc.63+ contributes to gastric cancer progression through regulation of NF-kB signaling. Gastric Cancer. 2020;**23**(5):863-873. DOI: 10.1007/s10120-020-01070-8

[68] Wojcik SE, Rossi S, Shimizu M, Nicoloso MS, Cimmino A, Alder H, et al. Non-codingRNA sequence variations in human chronic lymphocytic leukemia and colorectal cancer. Carcinogenesis. 2010;**31**(2):208-215. DOI: 10.1093/carcin/bgp209

[69] De Grassi A, Segala C, Iannelli F, Volorio S, Bertario L, Radice P, et al. Ultradeep sequencing of a human ultraconserved region reveals somatic and constitutional genomic instability. PLoS Biology. 2010;**8**(1):e1000275. DOI: 10.1371/journal.pbio.1000275

[70] Silla T, Kepp K, Tai ES, Goh L, Davila S, Catela Ivkovic T, et al. Allele frequencies of variants in ultra conserved elements identify selective pressure on transcription factor binding. PLoS One. 2014;**9**(11):e110692. DOI: 10.1371/journal.pone.0110692

[71] Halligan DL, Oliver F, Guthrie J, Stemshorn KC, Harr B, Keightley PD. Positive and negative selection in murine ultraconserved noncoding elements. Molecular Biology and Evolution. 2011;**28**(9):2651-2660. DOI: 10.1093/molbev/msr093

[72] Habic A, Mattick JS, Calin GA, Krese R, Konc J, Kunej T. Genetic variations of Ultraconserved elements in the human genome. OMICS. 2019;**23**(11):549-559. DOI: 10.1089/omi.2019.0156

[73] Katzman S, Kern AD, Bejerano G, Fewell G, Fulton L, Wilson RK, et al. Human genome ultraconserved elements are ultraselected. Science. 2007;**317**(5840):915. DOI: 10.1126/science.1142430

[74] Yang R, Frank B, Hemminki K, Bartram CR, Wappenschmidt B, Sutter C, et al. SNPs in ultraconserved elements and familial breast cancer risk. Carcinogenesis. 2008;**29**(2):351-355. DOI: 10.1093/carcin/bgm290

[75] Lin M, Eng C, Hawk ET, Huang M, Greisinger AJ, Gu J, et al. Genetic variants within ultraconserved elements and susceptibility to right- and left-sided colorectal adenocarcinoma. Carcinogenesis. 2012;**33**(4):841-847. DOI: 10.1093/carcin/bgs096

[76] Lin M, Eng C, Hawk ET, Huang M, Lin J, Gu J, et al. Identification of polymorphisms in ultraconserved elements associated with clinical outcomes in locally advanced colorectal adenocarcinoma. Cancer. 2012;**118**(24):6188-6198. DOI: 10.1002/cncr.27653

[77] Bao BY, Lin VC, Yu CC, Yin HL, Chang TY, Lu TL, et al. Genetic variants

in ultraconserved regions associate with prostate cancer recurrence and survival. Scientific Reports. 2016;**6**:22124. DOI: 10.1038/srep22124

[78] das Chagas PF, de Sousa GR, Kodama MH, CAO DBJ, Yunes JA, Brandalise SR, et al. Ultraconserved long non-coding RNA uc.112 is highly expressed in childhood T versus B-cell acute lymphoblastic leukemia. Hematology, Transfusion and Cell Therapy. 2021;**43**(1):28-34. DOI: 10.1016/j.htct.2019.12.003

[79] Sana J, Hankeova S, Svoboda M, Kiss I, Vyzula R, Slaby O. Expression levels of transcribed ultraconserved regions uc.73 and uc.388 are altered in colorectal cancer. Oncology. 2012;**82**(2):114-118. DOI: 10.1159/000336479

[80] Qian XX, Cai CW, Li HY, Lai LJ, Song DJ, Qiao YQ, et al. Transcribed ultraconserved region (T-UCR) uc.261 expression is closely correlated with disease activity and intestinal permeability in Crohn's disease. Therapeutic Advances in Gastroenterology. 2019;**12**:1756284819880733. DOI: 10.1177/1756284819880733

[81] Vannini I, Ferracin M, Fabbri F, Fabbri M. Overexpression of ultraconserved region 83- induces lung cancer tumorigenesis. PLoS One. 2022;**17**(1):e0261464. DOI: 10.1371/journal.pone.0261464

[82] Zadrożna-Nowak A, Romanowicz H, Zadrożny M, Bryś M, Forma E, Smolarz B. Analysis of long non-coding RNA (lncRNA) uc.38 and uc.63 expression in breast carcinoma patients. Genes (Basel). 2022;**13**(4):608. DOI: 10.3390/genes13040608

[83] Marini A, Lena AM, Panatta E, Ivan C, Han L, Liang H, et al. Ultraconserved long non-coding RNA uc.63 in breast cancer. Oncotarget. 2017;**8**(22):35669-35680. DOI: 10.18632/oncotarget.10572

[84] Sekino Y, Sakamoto N, Ishikawa A, Honma R, Shigematsu Y, Hayashi T, et al. Transcribed ultraconserved region uc.63+ promotes resistance to cisplatin through regulation of androgen receptor signaling in bladder cancer. Oncology Reports. 2019;**41**(5):3111-3118. DOI: 10.3892/or.2019.7039

[85] Liu X, Zhou X, Deng CJ, Zhao Y, Shen J, Wang Y, et al. Comprehensive analyses of T-UCR expression profiles and exploration of the efficacy of uc.63- and uc.280+ as biomarkers for lung cancer in Xuanwei, China. Pathology, Research and Practice. 2020;**216**(6):152978. DOI: 10.1016/j.prp.2020.152978

[86] Bomben R, Roisman A, D'Agaro T, Castellano G, Baumann T, Delgado J, et al. Expression of the transcribed ultraconserved region 70 and the related long non-coding RNA AC092652.2-202 has prognostic value in chronic lymphocytic leukaemia. British Journal of Haematology. 2019;**184**(6):1045-1050. DOI: 10.1111/bjh.15237

[87] Zambalde EP, Adamoski D, Gradia DF, Rabinovich I, Rodrigues AC, Ivan C, et al. Transcribed Ultraconserved Regions are associated with clinicopathological features in breast cancer. Biomolecules. 2022;**12**(2):214. DOI: 10.3390/biom12020214

[88] Ni W, Zhang Y, Zhan Z, Ye F, Liang Y, Huang J, et al. A novel lncRNA uc.134 represses hepatocellular carcinoma progression by inhibiting CUL4A-mediated ubiquitination of LATS1. Journal of Hematology & Oncology. 2017;**10**(1):91. DOI: 10.1186/

s13045-017-0449-4. Erratum in: J Hematol Oncol. 2023 May 8;16(1):51

[89] Kottorou AE, Antonacopoulou AG, Dimitrakopoulos FD, Diamantopoulou G, Sirinian C, Kalofonou M, et al. Deregulation of methylation of transcribed-ultra conserved regions in colorectal cancer and their value for detection of adenomas and adenocarcinomas. Oncotarget. 2018;**9**(30):21411-21428. DOI: 10.18632/oncotarget.25115

[90] Kottorou AE, Dimitrakopoulos FD, Antonacopoulou AG, Diamantopoulou G, Tsoumas D, Koutras A, et al. Differentially methylated ultra-conserved Regions Uc160 and Uc283 in adenomas and adenocarcinomas are associated with overall survival of colorectal cancer patients. Cancers (Basel). 2020;**12**(4):895. DOI: 10.3390/cancers12040895

[91] Wang L, Wang XC, Li X, Gu Y, Zhou J, Jiang S, et al. Expression of uc.189 and its clinicopathologic significance in gynecological cancers. Oncotarget. 2017;**9**(7):7453-7463. DOI: 10.18632/oncotarget.23761

[92] Guo Y, Wang C, Miao X, Chen S, Qian Y, Li G, et al. Upregulation of uc.189 in patients with esophageal squamous cell carcinoma and its clinicopathologic value. Pathology, Research and Practice. 2017;**213**(11):1400-1403. DOI: 10.1016/j.prp.2017.09.005

[93] Luo HL, Chen J, Luo T, Wu FX, Liu JJ, Wang HF, et al. Downregulation of macrophage-derived T-UCR uc.306 associates with poor prognosis in hepatocellular carcinoma. Cellular Physiology and Biochemistry. 2017;**42**(4):1526-1539. DOI: 10.1159/000479269

[94] Tian Y, Feng Y. Up-regulation of long noncoding RNA uc.338 predicts poor survival in non-small cell lung cancer. Cancer Biomarkers. 2018;**22**(4):781-785. DOI: 10.3233/CBM-181331

[95] Khalafiyan A, Emadi-Baygi M, Wolfien M, Salehzadeh-Yazdi A, Nikpour P. Construction of a three-component regulatory network of transcribed ultraconserved regions for the identification of prognostic biomarkers in gastric cancer. Journal of Cellular Biochemistry. 2023;**124**(3):396-408. DOI: 10.1002/jcb.30373

[96] Fassan M, Dall'Olmo L, Galasso M, Braconi C, Pizzi M, Realdon S, et al. Transcribed ultraconserved noncoding RNAs (T-UCR) are involved in Barrett's esophagus carcinogenesis. Oncotarget. 2014;**5**(16):7162-7171. DOI: 10.18632/oncotarget.2249

[97] Ciaramella A, Di Nardo E, Terracciano D, Conte L, Febbraio F, Cimmino A. A new biomarker panel of ultraconserved long non-coding RNAs for bladder cancer prognosis by a machine learning based methodology. BMC Bioinformatics. 2023;**23**(Suppl. 6):569. DOI: 10.1186/s12859-023-05167-6

Chapter 3

LncRNAs in Cancer Development

Alisa Petkevich, Aleksandr Abramov and Vadim Pospelov

Abstract

The goal of this chapter is to make an overview of the identified changes in lncRNAs expression levels accompanying cancer development. In general, the statistics allows us to establish a fact of association between the phenomenon and the process, but not to discover underlying mechanisms. In the context of the discussed topic, the phenomenon and the process are changes in lncRNA expression levels and cancer development. However, the underlying mechanisms, allowing such changes as in lncRNAs expression levels to have an impact on the cancer development, mostly remain uncertain and not clear. The first part of the chapter aims to shortly highlight the possible mechanisms of lncRNAs' impacts on the main processes of cancer development, like EMT, cancer cell progression, invasion, and metastasis. The second part examines in more detail the role of lncRNAs in some of the main nosology of oncological diseases: The participation of lncRNAs in the formation and development of breast cancer is considered, the main aspects of the importance of lncRNAs in lung cancer are presented, and studies on the participation of lncRNAs in the formation of colorectal cancer are described.

Keywords: lncRNAs, cancer development, breast cancer, lung cancer, colorectal cancer

1. Introduction

The protein-coding part of genomes, and the human genome is no exception, is not the least, but quantitatively an insignificant part of it: About 1.9% of transcribed RNA is responsible for protein coding, and in the mammalian genome, only 70% of RNA is transcribed [1]. A significant part of RNA is not protein coding, which raises the question of its role in the human beings and alive organisms in general. These non-coding RNAs are represented by molecules of different sizes and are classified according to the number of nucleotides and not according to their functional properties [2].

For a non-coding RNA of approximately 200 nucleotides in size, recent researches have elucidated many functions and mechanisms. It was experimentally demonstrated that the functions of such RNAs are associated with the mechanisms of DNA regulation, with epigenetic control of gene expression. These RNAs are involved in RNA maturation, and indirectly in gene regulation, for example, through interaction with promoters' sites. The following functions of non-coding RNAs have been identified: the inactivation of the X9 chromosome, formation and maintenance of nuclear architecture, and its importance in imprinting [3]. Small nuclear RNAs (snRNAs), as studies have shown, are involved in splicing processes, while transport RNA is responsible for the transport and identification of triplet nucleotides. Ribosomal RNAs, being the most represented in cells, are involved in the processes of formation of the ribosomal framework. Small nuclear RNAs (snoRNAs) perform, for example, functions related to housekeeping RNA chemical transformation processes [3]. In addition, there are many types of small non-coding RNAs involved in gene regulation, like miRNAs, which are 20–25 nucleotides and directly or indirectly regulate gene expression. Among these molecules are piRNAs (piwi-interacting RNAs), which differ from miRNAs in biogenesis and are mainly responsible for the targeting and switching off the expression of transposable and repetitive elements to support and maintain the constancy and stability of the embryonic genome [4]. The last most studied are siRNAs, which are involved in such processes as chromatin organization, control of mobile genetic elements, and gene regulation at the translation and at the post-translational level, and are also associated with the function of maintaining genome stability in some cells [4, 5].

Long non-coding RNAs (here and further lncRNAs) often contain a poly-A tail and can also be spliced, just like mRNA. It is worth noting that for a large number of lncRNAs, the functional significance has not yet been determined (and the exact genomic annotations have also not been identified). It is estimated that such lncRNAs constitute between 5400 and 10,000 of the lncRNA transcript pool. But still, the identification of mutations in the genes responsible for encoding those lncRNAs that are involved in the formation of cancer has made it possible to significantly rethink the molecular genetics of cancer development [6, 7].

In general, long and small non-coding RNAs are involved in the development of various diseases, including the development of cancer. lncRNAs are involved in processes such as angiogenesis, differentiation, and cell proliferation. They are also involved in apoptosis and cell migration, and are involved in the process of metastasis. lncRNAs can act both as oncogenes and as tumor suppressors [6, 7].

2. Long non-coding RNAs in cancer

Obviously, cancer is a complicated process involving many mechanisms and metamorphizing cell cycle in different steps. LncRNAs, which are expressed in various cell states, regulate gene expression processes so they may reflect the development of cancer. One possible mechanism is the binding of lncRNAs to proteins and the formation of scaffolds that are involved in the construction of regulatory complexes. One example of the functioning of such complexes is HOTAIR; it is involved in the trimethylation of LSD1, as well as H3K27, due to which demethylation of H3K4me2 occurs. HOTAIR implements this function through interaction with the PRC2 complex [8]. Qian et al. admit two other possible mechanisms of gene expression regulation: The first is the functioning of lncRNA as a decoy, which prevents the binding of transcription regulators. According to this theory, lncRNAs may bind miRNAs and prevent RNA degradation competing endogenous RNAs. The second possible way is guiding the transcription factors to specific sites like MEG3, which guides PRC2 and forms a complex with DNA [9].

Epithelial-mesenchymal transition or EMT is one of the key processes of cancer formation in epithelial cells and besides all cell types, while the main event of EMT is the total imbalance of epithelial cells following loss of the corresponding architecture, so anterior polarity and posterior polarity arise, apical-basal polarity is lost, and the shape of cells changes, what partly happen due to the damage to the structure of the cytoskeleton. These hallmark events are accompanied with disruption of epithelial gene expression markers that allow the determination of mesenchymal phenotypes along with cell growth and movement function changes. The new phenotype they acquire is often characterized with the possibility of extracellular matrix (ECM) protein degradation and apoptosis and aging ignoring [10]. All these events are realized through a majority of different molecular mechanisms, as it was already mentioned above, including disruption of signal pathways and changes in gene expression and regulation, and lncRNAs in this sense may play a pivotal role in the processes underlying EMT as well as in promotion cell invasion and migration. When examining lncRNAs in detail, it can be seen that a set of molecular processes occurring around them can significantly influence the functions of lncRNAs. In addition, depending on these processes, the functions of lncRNAs can become completely opposite. Liu et al. highlight the inconsistency of functions of H19, one of the earliest described lncRNAs63. In bladder cancer cells, it associates with the polycomb repressive complex 2 (PRC2) component EZH2, leading to decreased E-cadherin by increasing the activity of β-catenin, supporting the migration of malignant cells *in vitro* and a transition toward a mesenchymal state. Simultaneously, Liu mentioned studies in which H19 interacted with miR-200c, as well as h miR-200b. As a result of this interaction, GIT2 is suppressed, which in turn stimulates the formation of metastatic colonies. At the same time, in circulating tumor cells, H19 interacts with microRNA let-7b and blocks it, resulting in the suppression of CYTH3. Suppression of CYTH3 promotes the formation of a mesenchymal phenotype [11]. Along with this, lncRNAs are involved in the promotion of invasion and migration of tumor cells, so NORAD decrease has a significant correlation with lymph node metastases from patients with lung and breast cancer, while ectopic expression of this lncRNA *in vitro* inhibited invasion and migration [12]. The main mechanisms of action of this lncRNA on carcinogenesis and the development of metastases include suppression of S100P functions, due to which S100P-mediated signaling, which activates invasion and formation of metastases, is disrupted [12].

LncRNA CCAT2, through interaction with TCF7L2, activates MYC transcription and WNT signaling. Increased expression of TCF7L2 occurs in microsatellite-stable and metastatic CRC [13]. Ling showed that miR-20a and miR-17-5p are also activated by CCAT2 through TCF7L2-mediated transcriptional regulation, with CCAT2 being a downstream target of WNT, which may support a feedback mechanism. In addition, CCAT2 expression has been shown to be influenced by SNP status: The G risk allele produces more CCAT2 transcripts [13].

Jessica Silver-Fisher et al. found 148 differentially expressed RNAs associated with metastasis (RAMS). The results were obtained by sequencing the transcriptome of normal tissues, primary cancer tissues, and distant metastases. The main focus was on RAMS11 as it was most associated with poor disease-free survival. Additionally, RAMS11 has been shown to promote aggressive phenotypes in both *in vitro* and *in vivo* studies. The experiments presented in the work revealed the mechanism of RAMS11-dependent recruitment of Chromobox protein 4 (CBX4)—this protein is an activator of topoisomerase 2 alpha (TOP2α). RAMS11 expression was found to be highly overexpressed in liver metastases from colorectal cancer patients compared with its expression in primary tumor cells [14].

LncRNAs may participate in the activation of proliferation and processes underlying insensitivity to growth factors. LncRNAs may be involved in different mechanisms of proliferation activation such as activating the signal receptors or like PVT1 being involved in the regulation of receptor abundance. Some lncRNAs are involved in providing resistance to expression suppressors and tumor growth inhibitors by affecting different stages of translation and transcription. LincRNA-p21 and p21 (inhibitor of CDK2) are involved in the initiation of transcription by the scaffolding of transcription factor complexes, and gadd7 and Cdk6 participate in the destabilization of mRNA transcripts and thus modify transcription elongation [15].

Due to new technologies development, it became evident many lncRNAs are connected to carcinogenesis, making them new perspective players into the field of molecular biology. The main function of lncRNA is to regulate the expression of other genes, which gives these molecules a critical role in the existence of cells. Undoubtedly, the discovery of new biomarkers that allow the assessment of various aspects of oncological diseases and open up new opportunities for diagnosis, prognosis, and treatment is the most important task of modern medicine. In this regard, the study of molecules such as lncRNAs can be a promising approach. Next, we will focus and consider the involvement of lncRNAs in the development of breast cancer, lung cancer, and colorectal cancer.

3. LncRNA in breast cancer

Currently, breast cancer is the second leading cause of death among women worldwide, while being the most common cancer [16]. Recent studies have shown that lncRNA expression changes are associated with disease progression and survival in breast cancer patients [17–20]. Xu et al. showed that lncRNA 179 and lncRNA 216 have an association with the overall survival of breast cancer patients (P values <0.05 for both). In their study, they demonstrated significant changes in the expression levels of 10 lncRNAs associated with overall survival. The work included the following lncRNAs: MAPT-AS1, RP11-120 K18.2, SPACA6P-A, AC025016.1, LINC02037, SPACA6P-AS, RP11-616 M22.1, 7, RP1-37C10.3, RP11-454P21.1, and P11-344E13.4. To identify the relationship with the survival of breast cancer patients, the researchers conducted a univariate Cox regression analysis, which showed for 6 lncRNAs a higher regression coefficient and a decrease in overall survival. A protective role was revealed for lncRNA MAPT-AS1—the expression level of this lncRNA correlated with survival rate in breast cancer patients [17]. In 2021, Qiao et al. demonstrated the effect of lncRNA408 on breast cancer progression, which was significantly enhanced in breast cancer cells undergoing EMT and in breast cancer tumor cells with lymph node metastasis compared to samples without lymph node metastasis [18]. Since lncRNA-408 acts as a sponge for miR-654-5p, it may be partially involved in mitigating miR-654-5p repression of its target LIMK1. This molecular mechanism is partially responsible for the development of breast cancer cells and their metabolic changes, since LIMK1, a gene involved in cytoskeleton stabilization, is one of the target genes of miR-654-5p. In addition, due to a decrease in the expression level of miR-654-5p, the expression of COL1A1, as well as MMP2 and ITGB1, increases [18].

Another lncRNA involved in the development of cancer, especially breast cancer, is the ST7-AS1 lncRNA. In addition to stimulating breast cancer, it may influence the development and progression of laryngeal squamous cell carcinoma. This is due to the involvement of the following molecular mechanisms that control the IL6/JAK STAT3 signaling pathway. This signaling pathway is generally associated with poor prognosis and is frequently activated in many cancers [19]. Reduced expression of lncRNA ST7-AS1 in breast cancer correlates with clinical and advanced pathological characteristics (high grade, HER2 status, age, advanced stage, menopausal stage, histology), poor prognosis, and survival time. GO analysis showed that lncRNA ST7-AS1 is associated with T helper cells and DC cells [20].

Yi et al. found the influence of SLC26A4-AS1 on breast cancer development: In breast cancer patients, decreased SLC26A4-AS1 expression correlated with tarragon receptor status, age, and menopausal status. Low expression was also associated with poor overall survival. SLC26A4-AS1 has been shown to be involved in multiple biological processes required for cancer development, such as protein export, cristae formation, the endosomal sorting complex required for histone modification and transport, ATP generation, and others [21].

DRAIC, also known as LOC145837 and RP11-279F6.1, along with breast cancer has been shown to affect both lung cancer and prostate cancer. Briefly, the expression of this lncRNA in tumor cells was identified as estrogen-dependent and elevated relative to that of neighboring non-cancer cells. Zhao et al., working with RNA sequencing data from the Tumor Genome Atlas, analyzed more than 800 breast cancer samples compared with about 100 normal cell samples and found that the expression level of lncRNA DRAIC was associated with progesterone and estrogen receptors, and also with human epidermal cells [22].

Fonseca-Montaño et al. worked with the BRCA-TCGA and GEO-GSE96058 datasets and found that in patients with luminal B breast cancer, LINC00426 expression levels correlated with overall survival. Further studies showed that LINC00426 expression is associated with various oncological processes and immune pathways in breast cancer. In addition, the expression of LINC00426 was associated with the expression of genes with cytolytic activity, the expression of immune checkpoints, and the level of infiltration of various populations of immune cells [23].

Collina et al., working with biological samples rather than ready-made datasets, found that lncRNA HOTAIR, one of the first to be linked to cancer, was implicated in breast cancer development. Experiments have demonstrated that the expression level of lncRNA HOTAIR is associated with metastases to lymph nodes. In addition, the study revealed that the expression levels of HOTAIR lncRNA have a direct correlation with the expression levels of androgen receptors in samples obtained from patients with triple-negative breast cancer. Based on the data obtained in this way, one can assume a prognostic role of HOTAIR in TNBC, as well as its possible indirect participation in the regulation of the androgen receptor pathway [24].

Gene expression is regulated by lncRNA through different molecular mechanisms. Among them are chromatin modification, translation, and transcription. All these make LncRNA an important biomarker of the oncological process, reflecting those properties of tumor cells that allow them to metastasize, form drug resistance, acquire the ability to invade, vascularize, and progress.

A year before, Li et al. demonstrated, that another lncRNA, lncATB, was shown to be involved in breast cancer development and to be increased in breast cancer cells. The results of experiments on lncATB made it possible to assume its significant contribution to oncological processes, such as invasion, cell migration, and the ability to form clonal cells. Data were obtained from both *in vitro* and *in vivo* studies. lncATB has the ability to bind microRNAs of the miR-200 family, due to which it increases the expression of Twist 1. This, in turn, promotes epithelial-mesenchymal transition [25]. All these have shown that the lncATB expression has a direct positive correlation with tumor size and an inverse correlation with the duration of relapse-free survival and overall survival [25].

These lncRNAs, involved in breast cancer development, are listed below in **Table 1** and are shown in **Figure 1**.

#	LNC	Materials	Methods	Prognostic/ diagnostic value	Reference
1	RP11-616 M22.1, LINC02037, RP11-344E13.4, RP11-454P21.1, SPACA6P-AS, MAPT-AS, RP1-37C10.3	Biomaterial under study: breast cancer cells—913, non-tumor cells adjacent to cancer—cancer cells—113. Data: clinical data, training data (n = 608), test data set (n = 305).	Illumina HiSeq RNA-Seq	For a panel of 7 lncRNA markers, the ROC curve area was 0.771 for predicting 3-year survival. For a panel of 7 lncRNA markers, the ROC curve area was 0.780 for predicting 5-year survival.	[17]
2	LNC408	BC tissues in patients with BC and lymph node metastasis (N = 35) and with BC and without lymph node metastasis (N = 25) (stage I (N = 19), stage II (N = 18), stage III (N = 23)).	qRT-PCR	Lnc-408 was significantly increased in BC tissues with lymph nodes compared with BC tissues without lymph nodes ($P < 0.0001$). Lnc-408 expression level was directly related to N stage, TNM stage, and histological grade.	[18]
3	LncRNA ST7-AS1	15 PCR samples; clinical data on breast cancer from the TCGA project; 1065 samples HTSeq-Counts and HTSeq-FPKM.	RT-PCR. TCGA data	Compared with 1065 BC tissues, lncRNA ST7-AS1 expression was significantly higher in 111 healthy cells ($p < 0.001$). The low expression of lncRNA ST7-AS1 in breast cancer was correlated with a number of clinical characteristics: age, menopause, HER2 status, stage, histological type, survival, and poor prognosis.	[20]
4	SLC26A4-AS1	1109 tissues from patients with breast cancer; 113 normal breast tissue; clinical and RNAseq data from the TCGA project	Single-gene differential analysis of SLC26A4-AS1 and GSEA	The expression level of SLC26A4-AS1 in breast cancer tissues is lower than in normal breast tissues (0.267 ± 0.015 vs. 0.808 ± 0.062, $P < 0.001$, AUC was 0.805). SLC26A4-AS1 expression correlated with menopausal status, ER status, PR, and age.	[21]

#	LNC	Materials	Methods	Prognostic/ diagnostic value	Reference
5	*DRAIC* (also known as *LOC145837* and *RP11-279F6.1)*	828 tumor tissues and 105 normal breast tissues from TCGA		In patients with advanced stages of cancer, the expression of lncRNA DRAIC is higher than in patients with early stages. In patients with lymph node metastases, the expression of lncRNA DRAIC is higher than in patients without lymph node metastases. Higher *DRAIC* predicts shorter overall survival of patients.	[22]
6	*LINC00426*	BRCA-TCGA (TCGA database through cBioPortal) and GEO-GSE96058 (GDC Data Portal datasets). A total of 927 patients with breast cancer: 490 patients with luminal type A breast cancer; 192 patients with breast cancer luminal type B; 77 patients with HER2-positive breast cancer; 168 patients with bilateral breast cancer. A total of 3052 patients—validation group from the GEO database: 1657 patients with luminal type A breast cancer; 729 patients with breast cancer luminal type B; 327 patients with HER2-positive breast cancer; 339 patients with bilateral breast cancer.	Comparison of gene expression between groups of breast cancer patients with high and low expression of LINC00426 was performed using the R package DESeq2 (version 1.38.1). Values less than 10 were filtered out. RNA-seq data from breast cancer patient cohorts with high and low LINC00426 expression were assessed using GSEA (version 4.1.0).	Patients with non-luminal breast cancer showed increased expression of LINC00426 compared to patients with luminal breast cancer. It was also noted that LINC00426 expression has a prognostic value for overall survival in patients with luminal subtype B breast cancer with PAM50 from the TCGA cohort.	[23]

#	LNC	Materials	Methods	Prognostic/ diagnostic value	Reference
7	LncRNA HOTAIR	A total of 163 patients who underwent a mastectomy, quadrantectomy, or metastectomy.	HOTAIR *in situ* was determined using RNAscope.	HOTAIR lncRNA expression is associated with lymph node metastasis and ductal histological type. No association was found between lncRNA HOTAIR and tumor size, proliferation, distant metastases, and overall and disease-free survival.	[24]
8	LncRNA ATB	A total of 131 samples from patients with primary breast cancer following surgery. Sixteen samples of normal breast tissue. None of the patients received radiation therapy, endocrine therapy, chemotherapy, or targeted therapy.	Expression analysis of lncRNA EMT and lncATB was performed using real-time PCR (the study was performed in triplicate using specific primers). Normalization was performed using the 2-ΔΔCt method; the expression level of β-actin was used for normalization. To investigate the molecular mechanism, biotin-labeled lncATB probes were used and incubated with cytoplasmic lysate of the MCF-7 cell line. The resulting lncRNA-miRNA complexes were then collected using specific magnetic beads. qRT-PCR was used to detect miR-200.	In cells of triple-negative and luminal breast cancer B (Her-2-positive), there are increased levels of expression of lncRNAs LncATB and Twist1. Luminal type A breast cancer cells have lower expression levels of lncRNAs LncATB and Twist1. In normal breast cells, the expression levels of lncRNAs LncATB and Twist1 are the lowest. For lncATB, the AUC was 0.851. For Twist1, the AUC was 0.778. OS and DFS are longer in the low lncATB expression group than in the high lncATB expression group. The binding of lncATB to the miR-200 family has been demonstrated.	[25]

Table 1.
LncRNA and breast cancer.

4. LncRNA in lung cancer

LncRNAs, as in other cancers, play a significant role in lung cancer. LncRNAs PVT1 and also MALAT1 can be noted—their expression is often increased in lung tumors. Blocking these lncRNAs, or vice versa, activation, affects proliferation and cell growth [26]. LINC00680 is another lncRNA implicated in the development of

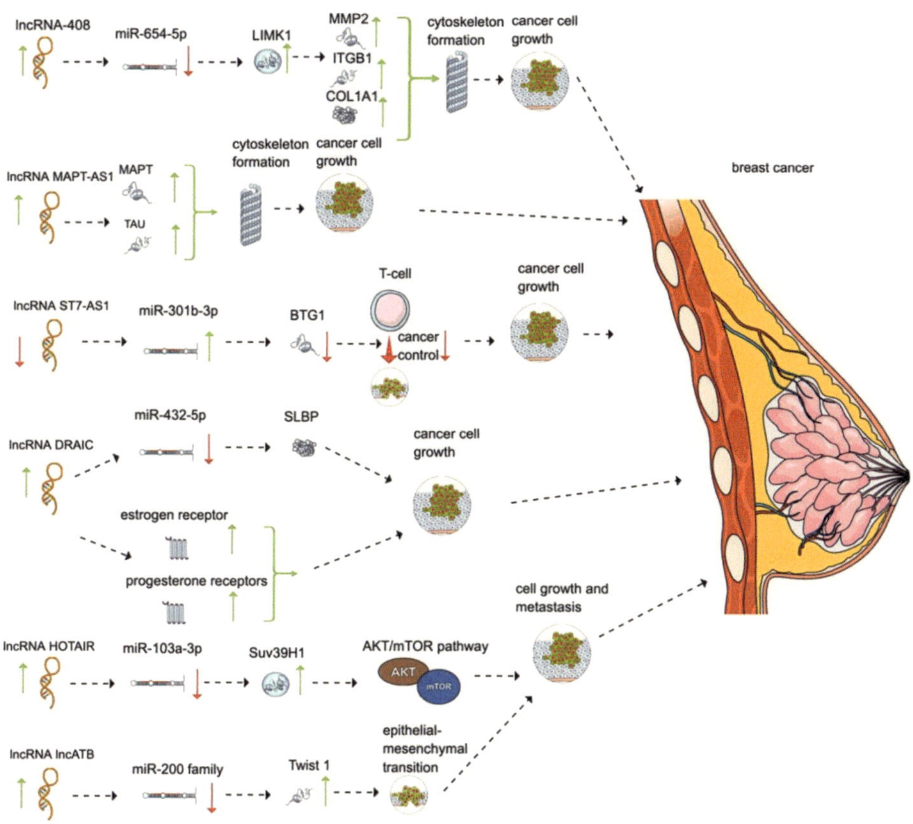

Figure 1.
LncRNAs in breast cancer development.

lung cancer. This lncRNA binds to GATA6, LINC00511, and the chromatin-converting enzyme EZH2, thereby suppressing the tumor suppressor genes p57 and LATS2, resulting in the development of non-small-cell lung cancer [26].

Hu et al. in 2016 demonstrated a promising potential of lncRNAs as non-invasive markers in lung cancer: They showed a high diagnostic value of these molecules working with plasma lncRNAs. From 21 previously identified lncRNAs, they identified SPRY4-IT1, ANRIL, and NEAT1 as the targets to be validated by qRT-PCR. The lncRNAs SPRY4-IT1, ANRIL, and NEAT1 were found to be significantly increased in plasma samples from patients with non-small-cell lung cancer. At the same time, ANRIL in the blood plasma of such patients proved to be the most informative diagnostic marker (ROC = 0.798). At the same time, when combined with lncRNA SPRY4-IT1 and NEAT1, the diagnostic significance increases (specificity 92.3%, sensitivity 82.8%) [27]. The training set included an independent group of 20 plasma samples from patients, and the validation set included 50 NSCLC patients and 50 different independent groups, allowing study control to confirm apparent differences in relative expression levels of biomarker candidates and 50 healthy controls, respectively [27].

Esfandi et al. obtained the data for NEAT1 expression in lung cancer with same tendencies as in the study by Hu et al. mentioned above. However, they worked with lncRNAs isolated from tissue samples, not from plasma, and showed that lncRNAs such as PVT1, HOXA transcript antisense RNA myeloid-specific 1 (HOTAIRM1), Fas-antisense 1 (FAS-AS1), taurine-regulated gene 1 (TUG1), Growth arrest-specific 5 (GAS5), copy nuclear paraspeckle accumulation 1 (NEAT1), TNFα, and hnRNPL-related immunoregulatory LincRNA (THRIL) have diagnostic and prognostic potential. Changes in their expression levels were shown by Esfandi et al. in 32 NSCLC samples and their corresponding non-cancerous cells [28]. It is interesting to admit a gender-dependent expression in some lncRNAs: NEAT1 showed a significant increase in NSCLC cells in men compared to the corresponding non-cancerous cells, and this change was not repeated in women and these data were confirmed in other published data of Pan et al. in 2015 and Hu et al. in 2016. Conversely, in non-small-cell lung cancer cells from male and female patients, FAS-AS1 expression was significantly reduced compared to corresponding normal lung cells. The expression levels of HOTAIRM1, as well as the expression levels of GAS5, TUG1, and THRIL, are significantly reduced in non-small-cell lung cancer cells compared with normal lung cells from male patients.

Increased expression of lncRNA RMRP is also characteristic of lung cancer. Meng et al. analyzed lncRNA expression in lung adenocarcinoma samples as well as in adjacent normal tissue samples (25 of 35). Experiments showed that in 71% of cases in adenocarcinoma samples, the expression level of lncRNA RMRP was increased compared to normal tissues. RMRP expression was also found to be increased in lung adenocarcinoma cell lines compared to bronchial epithelial cell lines. A possible molecular mechanism of RMRP may be inhibition of miR-206. Experiments with the H1299 cell line showed that cell proliferation and migration induced by RMRP are suppressed when miR-206 expression is restored [29].

Ren et al. [20] showed not only the association between the expression level of lncRNA and cancer development but also the association of mutations in lncRNAs sequences and cancer development. HOTAIR, which is known to be increased in a variety of cancers, was investigated in association with three SNPs identified with the MassArray system and these SNPs were rs920778, rs1899663, and rs4759314. In these studies, Ren et al. found that these SNPs were not significantly correlated with either lung cancer type, disease severity, metastases to lymph nodes, or cancer stage. There were also no correlations between these SNPs and family history. There is little gender dependence: using gender stratification, and the AG genotype compared to the AA genotype for rs920778 and the AG + GG genotypes for rs920778 were shown to be protective factors against NSCLC in women. In the analysis of smoking, it was shown that the AG + GG genotype was a protective factor against NSCLC in non-smokers compared with the AA genotype rs920778. In case with rs1899663 and rs4759314 genotypes, no statistical differences were demonstrated in both gender and smoking stratifications [30].

Lin et al. in 2015 obtained the same data about ANRIL overexpression in lung cancer as in the aforesaid study by Hu et al. Lin et al. analyzed ANRIL expression in 87 NSCLC tissues and three lung cancer cell lines. In tissue samples from patients with non-small-cell lung cancer, the expression level of lncRNA ANRIL was increased compared to samples from adjacent non-tumor tissues. In addition, the expression level of lncRNA ANRIL was correlated with higher-stage and late lymph node metastasis. It is noted that lower overall survival is observed in groups with high expression of lncRNA ANRIL. In addition, when the expression of lncRNA ANRIL is blocked, the proliferation, migration, and invasion of cancer cells *in vitro* are observed [31].

These lncRNAs, involved in breast cancer development, are listed below in **Table 2** and in **Figure 2**.

#	LNC	Materials	Methods	Prognostic/diagnostic value	Reference
1	lncRNAs such as SPRY4-IT1, CCAT2, RGMB-AS1, ANRIL, NEAT, PCAT1.	Plasma samples NSCLC (n = 120), plasma healthy controls (n = 120).	qRT-PCR	SPRY4-IT1, ANRIL, NEAT were significantly increased in NSCLC compared to samples obtained from normal lung cells.	[27]
2	FAS-AS1, GAS5, HOTAIRM1, NEAT1, TUG1, and TNFα - associated with hnRNPL	32 tissue samples from patients with non-small-cell lung cancer. No patient received either radiotherapy or chemotherapy.	Quantitative PCR. ANCT expression level was used for normalization. Expression was assessed by Efficiency^ΔCT.	Expression levels of HOTAIRM1, GAS5, TUG1, and THRIL, normalized by ANCT expression, are increased in tissue samples from men with NSCLC compared with normal tissues. The expression level of lncRNA FAS-AS1, normalized by ANCT expression, is significantly reduced in tissue samples from women with NSCLC compared with normal tissues. The expression level of lncRNA FAS-AS1, normalized by ANCT expression, is significantly reduced in tissue samples from men with NSCLC compared with normal tissues. The expression level of lncRNA NEAT1, normalized by ANCT expression, does not change in tissue samples from women with NSCLC compared to normal tissues. The expression level of lncRNA NEAT1, normalized by ANCT expression, is significantly increased in tissue samples from men with NSCLC compared with normal tissues.	[28]

#	LNC	Materials	Methods	Prognostic/diagnostic value	Reference
3	LncRNA RMRP	35 lung cancer tissue samples. 35 normal tissue samples (taken next to lung cancer tissue).	Real-time PCR. mRNA expression was normalized to GAPDH and lncRNA expression was normalized to 18S rRNA.	In 25 of 35 lung adenocarcinoma tissue samples (71%), RMRP expression was increased. Ectopic expression of RMRP increased lung adenocarcinoma cell proliferation, colony formation, and invasion.	[29]
4	LncRNA HOTAIR	Blood samples from patients with lung cancer—196 samples. Blood samples from healthy individuals who visited the hospital during the same period but did not have cancer or lung disease—196 samples.	GEPIA (Gene Expression Profile Interactive Analysis) resource was used to evaluate expression. MassArray system was used for analysis of HOTAIR lncRNA SNPs (rs920778, rs4759314, and rs1899663). Multiplex PCR—used to analyze 10–20 ng of genomic DNA. The 384-element SpectroCHIP array was used for locus-specific single-base extension reactions.	HOTAIR expression did not show a significant correlation with overall survival of patients with lung adenocarcinoma. There was also no significant correlation between the level of HOTAIR expression and the prognosis of the disease in male patients with lung adenocarcinoma and squamous cell lung cancer, as well as in female patients with lung adenocarcinoma and squamous cell lung cancer.	[30]
5	lncRNA ANRIL	Tissue samples from patients with non-small-cell lung cancer—87 samples. Lung cancer cell lines are SPC-A1, A549, and NCI-H1650. The cell line of normal human bronchial epithelial cells is 16HBE.	qRT-PCR	In tissue samples from patients with non-small-cell lung cancer, the expression level of lncRNA ANRIL was increased compared to samples from adjacent non-tumor tissues. In addition, the expression level of lncRNA ANRIL was correlated with higher-stage and late lymph node metastasis.	[31]

Table 2.
LncRNA and lung cancer.

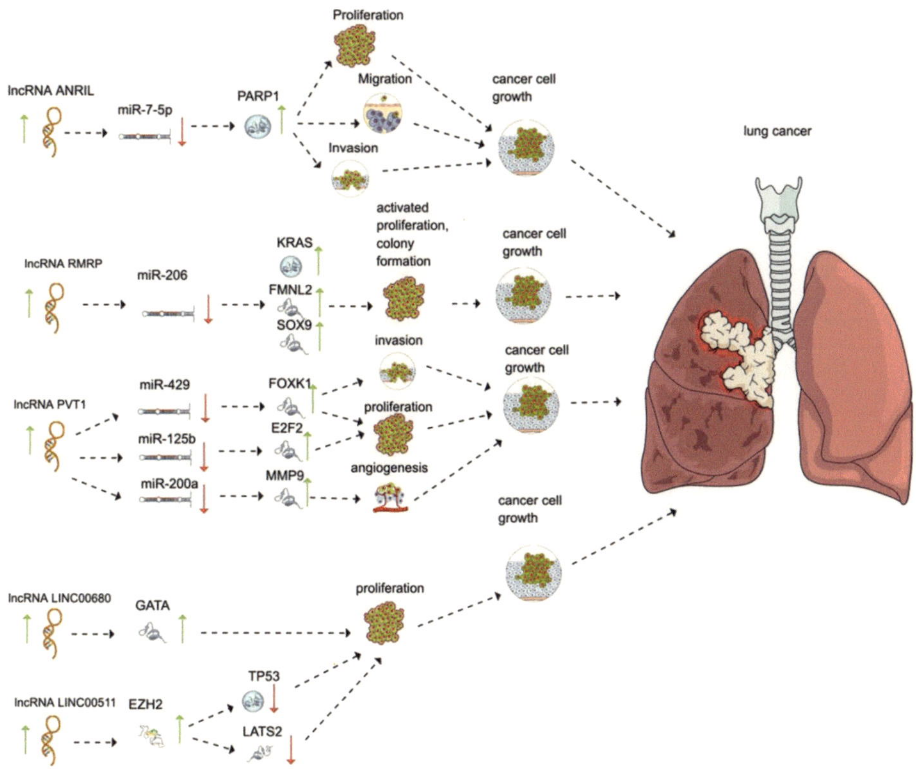

Figure 2.
LncRNAs in lung cancer development.

5. LncRNA in colorectal cancer

Ling et al. [13] and Nissan et al. [32] conducted a series of experiments and were among the first to obtain data demonstrating the general significance and possible effects of lncRNAs in colorectal cancer. LncRNA CCAT1, as well as lncRNA CCAT2, is located on chromosome 8q24 [32, 33]. Wu et al. in 2023 published a large review on the role of lncRNAs in colorectal cancer. This review cited data from many studies that showed that the expression levels of lncRNAs CCAT1 and CCAT2 are significantly increased in samples obtained from tumor tissues compared with the expression level in normal tissues. In addition, high expression levels of these lncRNAs were correlated with poor prognosis. In addition to these lncRNAs, HOTAIR, H19, and MALAT-1 are also significant for colorectal cancer [34].

Ye et al. showed that NALT1, which has been implicated in the occurrence of gastric cancer (GC), was overexpressed in 76 CRC patients ranging from stages I to IV. Increased NALT1 expression is associated with advanced-stage cancer. Inhibition of NALT1 reduced cell proliferation, invasion, and migration. On the contrary, when NALT1 expression increased, cell proliferation, invasion, and migration were activated both *in vitro* and *in vivo*. Studies have shown that the putative molecular mechanism of action of NALT1 is realized through the binding of miRNA-574-5p, due to which the expression of PEG10 is increased [35].

Another lncRNA with a role in colorectal cancer is BCYRN1. In addition, studies have shown that high BCYRN1 expression promotes metastasis in colorectal cancer patients and is also generally associated with poor prognosis. In an *in vitro* study, experiments also showed that downregulation of lncRNA BCYRN1 using si-BCYRN1 resulted in a significant reduction in invasion and migration [36].

Just like lncRNA-408 for miR-654-5p or lncRNA-ATB for miR-200 or H19 for miRNA let-7b, lncRNA cancer susceptibility 15 (CASC15) binds to miRNA-4310 acting like a sponge. lncRNA CASC15, due to the suppression of microRNA-4310,

#	LNC	Materials	Methods	Prognostic/diagnostic value	Reference
1	LncRNA NALT1	76 patients with colorectal cancer from stages I to IV. None of these patients received either radiation therapy or chemotherapy.	RT-qPCR—expression level was used for normalization GAPDH. The studies were carried out in triplets. Expression measured by the 2-ΔΔCt method.	High NALT1 expression correlated with advanced cancer stage. Suppression of NALT1 expression reduced cell proliferation, migration, and invasion. Increased expression of NALT1 resulted in increased proliferation, activation of cell migration, and invasion both *in vivo* and *in vitro*.	[35]
2	ncRNA BCYRN1	150 samples of tumor tissue from patients with colorectal cancer after surgery. 150 normal tissues (taken near tumor tissue) from patients with colorectal cancer after surgery.	qRT-PCR	In tissue samples from patients with colorectal cancer, the expression level of lncRNA BCYRN1 is significantly increased compared to adjacent normal tissues, which is also associated with a poor prognosis.	[36]
3	LncRNA CASC15	Samples of tumor tissue and adjacent normal tissue from patients with colorectal cancer after surgical treatment—93.	RT-qPCR, normalization was performed according to the level of GAPDH expression.	An increased level of CASC15 expression led to the activation of the Wnt signaling pathway through mira4310, resulting in an increase in the proliferation of colon cancer cells and increased metastasis.	[37]
4	LncRNA SURC	Samples of tumor tissue from patients with colorectal cancer who underwent surgery—150.	qPCR and FISH	A higher number of SURC-positive cells in colorectal cancer tissues compared with adjacent normal tissues were detected by FISH staining. SURC was significantly more expressed in stage II–III than in stage I patients.	[39]

Table 3.
LncRNA and colorectal cancer.

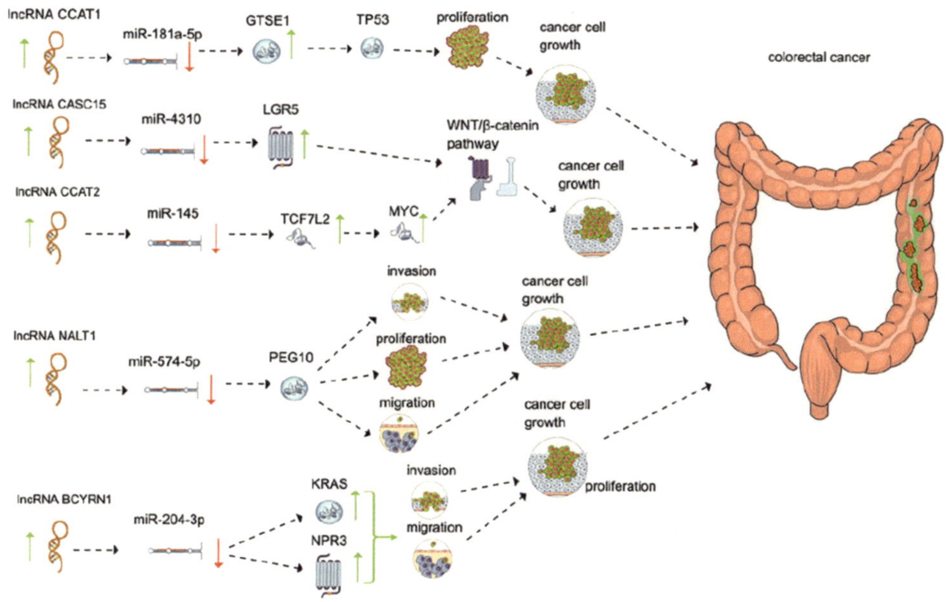

Figure 3.
LncRNAs in colorectal cancer development.

increases the expression of LGR5, due to which the Wnt signaling pathway is activated. Due to the activation of the Wnt signaling pathway, there is an increase in cell proliferation, invasion, and migration of colon cancer cells. It has also been demonstrated that CASC15 expression is increased in colorectal cancer cells and is associated with metastasis and clinical stage [37]. These are examples of the formation of the lncRNA-microRNA-mRNA signaling axis in which lncRNA 15 and SURC play an important role. The lncRNA-miRNA mechanism in colorectal cancer was demonstrated by the study by Yu et al., where novel lncRNA-miRNA mechanism in colorectal cancer was described and where it was shown that CCAT2 can selectively inhibit miR-145 growth by inhibiting pre-miR-145 by Dicer and its export to the cytoplasm [38].

As it was mentioned above, another lncRNA-miRNA axis in colorectal cancer, promoting tumor development, is formed by lncRNA SURC, which is overexpressed in this type of cancer. SURC promotes CCND2 expression by inhibiting the expression of miR–185-5p in colorectal cancer cells, thus being a specifically upregulated lncRNA in colorectal cancer cells. Along with this, Li et al. have recently demonstrated that the SURC/miR–185-5p/CCND2 axis may be targetable for colorectal cancer diagnosis and therapy [39]. Changes in SURC expression level were identified while comparing lncRNA expression profiles in cancer and normal tissues in mice models. The elevation of lncRNA-mAK028845 expression and its upregulation during colorectal cancer initiation were shown in AOM/DSS-induced CAC model. However, the use of this lncRNA as cancer-specific is problematic, since, as was later revealed, it is actively expressed in non-cancerous tissues of mice. Sequencing revealed that lncRNA AK028845 (named SURC) is located on chromosome 17 in humans, and it contains an exon of the KRT37 gene and a pseudogene KRT41 [39].

These lncRNAs, involved in breast cancer development, are listed below in **Table 3** and in **Figure 3**.

6. Conclusions

However, the extent and nature of lncRNAs promoting the interlocking pathological hallmarks in breast cancer, lung cancer, colorectal cancer, or any other type of cancer remain unclear. Despite the advantage of lncRNA as a diagnostic marker compared to protein-coding RNAs, which is in their own expression as it is a better indicator of the tumor status, there remain some limitations of the study of lncRNA expression in different cancers. These limitations include its retrospective nature and bioinformatics approach based on transcriptomic data, which requires validation of these findings through methodologies like multiplex immunofluorescence or flow cytometry. This validation process is not always possible to be performed because of the unavailability of biological samples, for example. Anyway, lncRNAs remain a promising approach for biomarker studies with further use in clinical routine practice in cancer diagnosis, monitoring, and treatment.

Conflict of interest

The authors declare no conflict of interest.

Author details

Alisa Petkevich*, Aleksandr Abramov and Vadim Pospelov
Scientific and Practical Center of Specialized Health Care for Children named after V.F. Voino-Yasenetskiy, Moscow, Russia

*Address all correspondence to: pa.alisa26@gmail.com

References

[1] Uchida S, Dimmeler S. Long noncoding RNAs in cardiovascular diseases. Circulation Research. 2015;**116**:737-750. DOI: 10.1161/CIRCRESAHA.116.302521

[2] Uchida S. High-throughput methods to detect long non-coding RNAs. High Throughput. 2017;**6**(3):12. DOI: 10.3390/ht6030012 [Accessed: August 31, 2017]

[3] Matera AG, Terns RM, Terns MP. Non-coding RNAs: Lessons from the small nuclear and small nucleolar RNAs. Nature Reviews. Molecular Cell Biology. 2007;**8**:209-220

[4] Hombach S, Kretz M. Non-coding RNAs: Classification, biology and functioning. Advances in Experimental Medicine and Biology. 2016;**937**:3-17. DOI: 10.1007/978-3-319-42059-2_1

[5] Piatek MJ, Werner A. Endogenous siRNAs: Regulators of internal affairs. Biochemical Society Transactions. 2014;**42**(4):1174-1179. DOI: 10.1042/BST20140068

[6] Mattick JS, Amaral PP, Carninci P, et al. Long non-coding RNAs: Definitions, functions, challenges and recommendations. Nature Reviews. Molecular Cell Biology. 2023;**24**:430-447. DOI: 10.1038/s41580-022-00566-8

[7] Lulli M, Napoli C, Landini I, Mini E, Lapucci A. Role of non-coding RNAs in colorectal cancer: Focus on long non-coding RNAs. International Journal of Molecular Sciences. 2022;**23**:13431. DOI: 10.3390/ijms232113431

[8] Zappulla DC, Cech TR. Yeast telomerase RNA: A flexible scaffold for protein subunits. Proceedings of the National Academy of Sciences of the United States of America. 2004;**101**:10024-10029. DOI: 10.1073/pnas.0403641101

[9] Qian Y, Shi L, Luo Z. Long non-coding RNAs in cancer: Implications for diagnosis, prognosis, and therapy. Frontiers in Medicine (Lausanne). 2020;**7**:612393. DOI: 10.3389/fmed.2020.612393 [Accessed: November 30, 2020]

[10] Lamouille S, Xu J, Derynck R. Molecular mechanisms of epithelial-mesenchymal transition. Nature Reviews. Molecular Cell Biology. 2014;**15**(3):178-196. DOI: 10.1038/nrm3758

[11] Liu SJ, Dang HX, Lim DA, Feng FY, Maher CA. Long noncoding RNAs in cancer metastasis. Nature Reviews. Cancer. 2021;**21**(7):446-460. DOI: 10.1038/s41568-021-00353-1

[12] Tan B-S et al. LncRNA NORAD is repressed by the YAP pathway and suppresses lung and breast cancer metastasis by sequestering S100P. Oncogene. 2019;**38**:5612-5626

[13] Ling H et al. CCAT2, a novel noncoding RNA mapping to 8q24, underlies metastatic progression and chromosomal instability in colon cancer. Genome Research. 2013;**23**:1446-1461

[14] Silva-Fisher JM, Dang HX, White NM, et al. Long non-coding RNA RAMS11 promotes metastatic colorectal cancer progression. Nature Communications. 2020;**11**(1):2156. DOI: 10.1038/s41467-020-15547-8 [Accessed: May 1, 2020]

[15] Bartonicek N, Maag JLV, Dinger ME. Long noncoding RNAs in cancer: Mechanisms of action and

technological advancements. Molecular Cancer. 2016;**15**:43. DOI: 10.1186/s12943-016-0530-6

[16] DeSantis CE, Ma J, Goding Sauer A, et al. Breast cancer statistics, 2017, racial disparity in mortality by state. CA: A Cancer Journal for Clinicians. 2017;**67**:439-448. DOI: 10.3322/caac.21412

[17] Xu M, Chen Z, Lin B, Zhang S, Qu J. A seven-lncRNA signature for predicting prognosis in breast carcinoma. Translational Cancer Research. 2021;**10**(9):4033-4046. DOI: 10.21037/tcr-21-747

[18] Qiao Y, Jin T, Guan S, et al. Long non-coding RNA Lnc-408 promotes invasion and metastasis of breast cancer cell by regulating LIMK1. Oncogene. 2021;**40**(24):4198-4213. DOI: 10.1038/s41388-021-01845-y

[19] Johnson DE, O'Keefe RA, Grandis JR. Targeting the IL-6/JAK/STAT3 signalling axis in cancer. Nature Reviews. Clinical Oncology. 2018;**15**:234-248. DOI: 10.1038/nrclinonc.2018.8

[20] Zhang Z, Zhang H, Li D, Zhou X, Wang J, Zhang Q. LncRNA ST7-AS1 is a potential novel biomarker and correlated with immune infiltrates for breast cancer. Frontiers in Molecular Biosciences. 2021;**8**:604261. Published 2021 Apr 12. DOI: 10.3389/fmolb.2021.60426

[21] Yi W, Shen H, Sun D, et al. Low expression of long noncoding RNA SLC26A4 antisense RNA 1 is an independent prognostic biomarker and correlate of immune infiltrates in breast cancer. Medical Science Monitor. 2021;**27**:e934522. DOI: 10.12659/MSM.934522 [Accessed: December 9, 2021]

[22] Zhao D, Dong JT. Upregulation of long non-coding RNA DRAIC correlates with adverse features of breast cancer. Noncoding RNA. 2018;**4**(4):39. DOI: 10.3390/ncrna4040039 [Accessed: December 11, 2018]

[23] Fonseca-Montaño MA, Cisneros-Villanueva M, Coales I, Hidalgo-Miranda A. LINC00426 is a potential immune phenotype-related biomarker and an overall survival predictor in PAM50 luminal B breast cancer. Frontiers in Genetics. 2023;**14**:1034569. DOI: 10.3389/fgene.2023.1034569 [Accessed: May 16, 2023]

[24] Collina F, Aquino G, Brogna M, et al. LncRNA HOTAIR up-regulation is strongly related with lymph nodes metastasis and LAR subtype of triple negative breast cancer. Journal of Cancer. 2019;**10**(9):2018-2024. DOI: 10.7150/jca.29670 [Accessed: May 12, 2019]

[25] Li RH, Chen M, Liu J, et al. Long noncoding RNA ATB promotes the epithelial–mesenchymal transition by upregulating the miR-200c/Twist1 axe and predicts poor prognosis in breast cancer. Cell Death & Disease. 2018;**9**:1171. DOI: 10.1038/s41419-018-1210-9

[26] Esposito R, Polidori T, Meise DF, et al. Multi-hallmark long noncoding RNA maps reveal non-small cell lung cancer vulnerabilities. Cell Genomics. 2022;**2**(9):100171. DOI: 10.1016/j.xgen.2022.100171 [Accessed: August 22, 2022]

[27] Hu X, Bao J, Wang Z, et al. The plasma lncRNA acting as fingerprint in non-small-cell lung cancer. Tumour Biology. 2016;**37**(3):3497-3504. DOI: 10.1007/s13277-015-4023-9

[28] Esfandi F, Taheri M, Omrani MD, et al. Expression of long non-coding RNAs (lncRNAs) has been dysregulated

in non-small cell lung cancer tissues. BMC Cancer. 2019;**19**(1):222. DOI: 10.1186/s12885-019-5435-5 [Assessed: March 12, 2019]

[29] Meng Q, Ren M, Li Y, Song X. LncRNA-RMRP acts as an oncogene in lung cancer. PLoS One. 2016;**11**(12):e0164845. DOI: 10.1371/journal.pone.0164845 [Assessed: December 1, 2016]

[30] Ren MM, Xu S, Wei YB, et al. Roles of HOTAIR in lung cancer susceptibility and prognosis. Molecular Genetics & Genomic Medicine. 2020;**8**(7):e1299. DOI: 10.1002/mgg3.1299

[31] Lin L, Gu ZT, Chen WH, Cao KJ. Increased expression of the long non-coding RNA ANRIL promotes lung cancer cell metastasis and correlates with poor prognosis. Diagnostic Pathology. 2015;**10**:14. DOI: 10.1186/s13000-015-0247-7 [Assessed: March 27, 2015]

[32] Nissan A, Stojadinovic A, Mitrani-Rosenbaum S, Halle D, Grinbaum R, Roistacher M, et al. Colon cancer associated transcript-1: A novel RNA expressed in malignant and pre-malignant human tissues. International Journal of Cancer. 2012;**130**(7):1598-1606

[33] Ling H, Spizzo R, Atlasi Y, Nicoloso M, Shimizu M, Redis RS, et al. CCAT2, a novel noncoding RNA mapping to 8q24, underlies metastatic progression and chromosomal instability in colon cancer. Genome Research. 2013;**23**(9):1446-1461

[34] Wu Y, Xu X. Long non-coding RNA signature in colorectal cancer: Research progression and clinical application. Cancer Cell International. 2023;**23**:28. DOI: 10.1186/s12935-023-02867-0

[35] Ye M, Zhao L, Zhang L, et al. LncRNA NALT1 promotes colorectal cancer progression via targeting PEG10 by sponging microRNA-574-5p. Cell Death & Disease. 2022;**13**(11):960. DOI: 10.1038/s41419-022-05404-5 [Assessed: November 16, 2022]

[36] Yu JH, Chen Y. Clinical significance of lncRNA BCYRN1 in colorectal cancer and its role in cell metastasis. European Review for Medical and Pharmacological Sciences. 2019;**23**(21):9371-9378. DOI: 10.26355/eurrev_201911_19430

[37] Jing N, Huang T, Guo H, et al. LncRNA CASC15 promotes colon cancer cell proliferation and metastasis by regulating the miR-4310/LGR5/Wnt/β-catenin signaling pathway. Molecular Medicine Reports. 2018;**18**(2):2269-2276. DOI: 10.3892/mmr.2018.9191

[38] Yu Y, Nangia-Makker P, Farhana L, Majumdar APN. A novel mechanism of lncRNA and miRNA interaction: CCAT2 regulates miR-145 expression by suppressing its maturation process in colon cancer cells. Molecular Cancer. 2017;**16**(1):155

[39] Li J, Ji Y, Chen N, et al. A specific upregulated long noncoding RNA in colorectal cancer promotes cancer progression. JCI Insight. 2022;**7**(15):e158855. DOI: 10.1172/jci.insight.158855 [Assessed: August 8, 2022]

Chapter 4

Circulating lncRNAs in Cancer

Le Qu, Cheng Chen, Meijuan Wu and Yulin Zhou

Abstract

Circulating lncRNAs (noncoding RNAs of 200 nucleotides or more) have been proposed as a new class of potential biomarkers for cancer diagnosis and prognosis. Circulating lncRNAs seem more reliable than other circulating nucleic acids due to their high stability in the blood. In this chapter, we will describe the characteristics of circulating lncRNAs; summarize the role of circulating lncRNAs as potential biomarkers for cancer diagnosis and cancer prognosis, respectively, focusing on their sensitivity and specificity as biomarkers; provide recommendations for the detection of circulating lncRNA levels; and finally discuss challenges encountered.

Keywords: circulating lncRNAs, cancer, biomarker, diagnosis, prognosis

1. Introduction

lncRNAs have been arbitrarily defined as non-coding transcripts greater than 200 nucleotides in length. Currently, up to 100,000 lncRNAs have been identified [1]. Due to lack of open reading frame (ORF), lncRNAs were regarded as unable to encode proteins [2]. However, a growing number of studies have currently confirmed that some lncRNAs containing short or small ORFs (smORFs) have the ability to encode functional polypeptides [3]. Like mRNAs, lncRNAs are transcribed by RNA polymerase II, spliced, 5′ capped, and 3′ polyadenylated [4]. But unlike mRNAs, many lncRNAs localize preferentially in the nucleus.

lncRNAs perform different functions according to their location in the cytoplasm or nucleus of cells. lncRNAs located in nucleus are involved in epigenetic and transcriptional modifications of different genes, such as histone modification, DNA methylation, and chromatin remodeling. lncRNAs can also interact with proteins and transcription factors in nucleus [5, 6]. However, some lncRNAs featured by 3′ cleavage and polyadenylation like mRNAs have the ability to transport to the cytoplasm [7], thereby modulating gene expression post-transcriptionally. They can act as competitive endogenous RNA (ceRNA) to interact with miRNAs, thereby affecting proteins in the cytoplasm and modulating RNA metabolism [6, 8, 9]. Increasing studies have confirmed diverse roles of lncRNAs in tumorigenesis and progression. They can act as either oncogenes or tumor suppressor genes [10, 11]. Their dysregulation contributes to various cancer-related events, such as cancer cell proliferation, migration, invasion, apoptosis, maintenance of stemness, and angiogenesis [12].

lncRNAs can be released into extracellular space and detected as circulating lncRNAs in body fluids such as plasma, serum, and urine [13]. They are highly stable in the bloodstream and poorly sensitive to nuclease-mediated degradation. It has been reported that neither repetitive freeze-thaw cycles nor prolonged exposure to 45°C and room temperature can induce the degradation of plasmatic lncRNAs [14]. The presence of extensive secondary structures [15] as well as stabilizing posttranscriptional modifications may account for the stability of lncRNAs in the bloodstream. Besides, some circulating lncRNAs may interact with circulatory proteins such as Argonaute (Ago) when transporting, protecting themselves from degradation [16].

Intriguingly, circulating lncRNAs can, at least for a part, be packaged into extracellular vesicles (EV) such as exosomes to transport in the blood [17, 18]. Notably, lncRNAs may be selectively sorted into exosomes due to the fact that some lncRNAs are enriched in exosomes, while others are barely present. Not merely as a carrier in extracellular space, exosomes also facilitate intercellular transferring of lncRNAs. Emerging evidence has shown that exosomal lncRNAs play vital roles in facilitating tumorigenesis by regulating angiogenesis, immunity, and metastasis [19].

Thus, due to the crucial roles of lncRNAs in cancer, along with the relatively easy and non-invasive detection in body fluids, circulating lncRNAs are emerging as superior biomarkers for cancer diagnosis and prognosis.

2. Circulating lncRNAs as potential biomarkers for cancer diagnosis

The feasibility of circulating lncRNAs in the diagnosis of cancers is currently a subject of increasing interest. In the following sections, we summarize circulating lncRNAs with diagnostic advantages over conventional cancer biomarkers, focus on the sensitivity and specificity of each single lncRNA in relevant cancers, and emphasize their promising potential in early cancer detection.

2.1 Breast cancer (BC)

Breast cancer (BC) is one of the most common cancers in the world and remains the leading cause of cancer-related deaths in women [20]. Delayed diagnosis, recurrence, and metastasis remain main obstacles for BC treatments [21]. Conventional serum biomarkers, such as CA125 and CEA, have limitations in clinical utility and remain controversial in BC patients [22, 23]. Recently, some lncRNAs have emerged as ideal biomarkers for BC diagnosis due to its ease of use and low invasiveness. For example, plasma levels of lncRNA H19 have been found significantly higher in women with BC than in healthy individuals. Moreover, compared with CA153 and CEA, circulating H19 had better diagnostic power (sensitivity 56.7%, specificity 86.7%, and AUC 0.81). The combination of three biomarkers can further elevate the diagnostic power of BC (sensitivity 56.7%, specificity 96.7%, and AUC 0.84) [24]. The higher performance of circulating H19 in BC diagnosis was also confirmed by another study, in which the AUC (0.870) for exosomal H19 was higher than the AUCs for CA153 (0.822) and CEA (0.811) [25].

Similarly, the levels of lncRNA HOTAIR in the plasma of BC patients were higher than in healthy individuals ($P < 0.05$). Compared with CA153 and CEA, circulating HOTAIR levels had better performance in diagnostic power (sensitivity 69.2%, specificity 93.3%, and AUC 0.80). The combination of three plasma biomarkers slightly increased the diagnostic power (sensitivity 73.1%, specificity 90%, and AUC 0.82) [26]

Another lncRNA which is detectable in the serum of BC patients is MALAT-1. The serum levels of MALAT-1 in BC patients are more than those in non-breast cancer patients with the sensitivity and specificity of 83.7% and 81.2%, respectively [27], which is better than CA153 and valuable for BC diagnosis. Likewise, lncRNA-ATB was significantly overexpressed in sera of BC patients compared to controls and had a higher AUC value than CA153 (AUC: 0.844, versus 0.73) for early diagnosis of BC patients with stage I-II [28].

2.2 Non-small-cell lung cancer (NSCLC)

Lung cancer (LC) remains the leading cause of cancer-related mortality worldwide [29]. Non-small-cell lung cancer (NSCLC) and small-cell lung cancer (SCLC) are two major subtypes of LC, and the former accounts for 80–85 % of all patients. Most NSCLC patients are diagnosed at advanced stages, and current biomarkers for NSCLC detection, such as CEA, NSE, TPA, chromogranin, CA125, CA199, and CYFRA211, remain highly sensitive but lowly specific [30–32]. NSCLC patients had higher plasma levels of HOTAIR than healthy individuals, which were decreased after tumor resection [33]. Moreover, circulating HOTAIR had higher diagnostic power than CEA (AUC 0.791 versus AUC 0.737), and the combination of two biomarkers can better discriminate NSCLC patients from healthy individuals (AUC 0.841) [33]. Likewise, Linc00152 levels were upregulated in both NSCLC tissues and plasma samples. Postoperative samples had significantly lower levels of plasma Linc00152 than preoperative samples. Their combination could provide more powerful diagnosis efficiency than Linc00152 or CEA alone (AUC 0.881) [34].

By contrast, the plasma levels of GAS5 were decreased in NSCLC patients compared with healthy individuals, which were increased following tumor resection [35]. Furthermore, circulating GAS5 can better distinguish NSCLC patients from healthy individuals (sensitivity 82.2%, specificity 72.7%, and AUC 0.832) compared with CEA levels, and the combination of two biomarkers further increased the diagnostic power (sensitivity 86.7%, specificity 90.9%, and AUC 0.909) [35]. In addition, the combination of GAS5, CEA, and CA199 had better diagnostic power than each biomarker individually [36].

A circulating 4-lncRNA panel (RMRP, NEAT1, TUG1, and MALAT1) had been identified to provide a high diagnostic value in NSCLC (AUC = 0.86 and 0.89 for training and verification phase, respectively). Compared with routine protein panels (CEA, CA125, and CYFRA21-1), the sensitivity of 4-lncRNA panel was higher (86.30% vs. 73.96%) in adenocarcinoma. Combination of 4-lncRNA with protein markers had higher diagnostic capacity in both adenocarcinoma (AUC = 0.85) and squamous cell carcinoma (AUC = 0.93) [37].

A recent study has reported that exosomal lncRNA RP5-977B1 exhibited higher levels in NSCLC than that in healthy controls. The AUC value of exosomal RP5-977B1 was higher than that of CYFRA21-1 and CEA in both testing and multicentric validation cohort [38]. Similarly, the level of lncRNA HEIH in peripheral blood of lung squamous cell carcinoma (LUSC)/lung adenocarcinoma (LUAD) patients was higher than that in healthy controls. lncRNA HEIH had higher diagnostic efficiency than CEA in both LUSC (AUC 0.860 vs 0.706, sensitivity 72.86% vs 31.43%, specificity 95.71% vs 100.00%) and LUAD patients (AUC 0.905 vs 0.763, sensitivity 77.50% vs 50.00%, specificity 95.71% vs 90.00%) [39].

2.3 Gastric cancer (GC)

Gastric cancer (GC) remains one of the most common malignancies and a leading cause of mortality worldwide [40]. Most GC patients are at an advanced stage when diagnosed [41]. Current diagnostic approaches, such as pathological biopsies or digestive tract endoscopy, mainly rely on invasive techniques and are hard to be applied for screening purposes. Recently, lncRNAs have emerged as noninvasive biomarkers, which play important roles in early diagnosis of GC [42]. For example, the expression levels of exosomal lncRNA HOTTIP were reported to be typically upregulated in GC than in normal control. Exosomal HOTTIP showed higher diagnostic capability than conventional biomarkers CEA, CA199, and CA724 (AUC 0.827 vs 0.653, 0.685 and 0.639, respectively) [43].

A recent research has demonstrated that lncRNA HULC and ZNFX1-AS1 can distinguish gastric cancer patients from healthy controls, with AUC value of 0.65 and 0.85, respectively. Furthermore, the AUC values of HULC and ZNFX1-AS1 were higher than those of traditional serum biomarkers, including CEA (0.62), CA199 (0.56), CY211 (0.59), and NSE (0.56) [44]. Another study further confirmed that HULC had higher sensitivity and specificity than CA724 and CEA in the diagnostic of GC [45]. Likewise, lncRNA B3GALT5-AS1 has been reported to be better than CEA and CA199 as a diagnostic biomarker of GC [46]. Both serum AFAP1-AS1 and FEZF1-AS1 were upregulated in GC patients compared with normal individuals, and the AUC value demonstrated the two lncRNAs had higher diagnostic utility than CEA and CA19-9 [47].

Circulating exosomal lncRNAs PCSK2-2:1 (AUC 0.9, sensitivity 84%, and specificity 86.5%) and GNAQ-6:1 (AUC 0.74, sensitivity 83.7%, and specificity 55.6%) have been reported as better biomarkers for distinguishing GC patients from healthy people when compared with traditional diagnostic biomarkers, such as CA724 (AUC 0.57, sensitivity 56%, and specificity 65.5%) [48, 49]. lnc-SLC2A12-10:1 was remarkably upregulated in exosomes derived from patients with GC with AUC value of 0.776, which was higher than the diagnostic accuracies of CEA, CA 19-9, and CA72-4 [50].

A recent study has demonstrated that lncRNA-GC1 had better diagnostic performance than standard biomarkers CEA, CA19-9, and CA72-4 in distinguishing GC patients from healthy individuals. Moreover, lncRNA-GC1 showed superiority in early detection of GC, especially for patients with negative standard biomarkers [51].

Intriguingly, lncRNAs combined with traditional serum tumor markers have improved the diagnostic efficiency of GC. For example, lncRNA CTC-497E21.4 combined with CEA and CA199 can increase the AUC value to 0.9 [52]. Likewise, the combination of lncRNA HCP5 with CEA and CA199 can increase the AUC value, sensitivity, and specificity to 81%, 79%, and 0.80, respectively, when diagnosing GC [53]. Similarly, lncRNA FAM49B-AS, GUSBP11, and CTDHUT combined with CA242 and CA724 can increase the AUC value, sensitivity, and specificity to 95%, 93.2%, and 0.866, respectively [54]. The level of lncRNA RP11-731F5.2 was significantly higher in the serum of GC patients than in healthy controls and decreased subsequently after surgery. RP11-731F5.2 combined with CEA and CA19-9 had the highest AUC value of 0.84, which significantly improved the diagnostic efficacy of GC [55].

Gastric juice is an ideal sample for studying GC as it is directly secreted by gastric mucosa and reflect the pathological state of the stomach [56]. Studies have revealed that lncRNAs can be identified from gastric juice and used for GC diagnosis, for example, RMRP, AA174084, PVT1, H19, LINC00982, ABHD11-AS1, UCA1, and LINC00152. The sensitivities, specificities, and AUC values of these lncRNAs range from 41 to 56.4%, 75.4 to 93.4%, and 0.65 to 0.85, respectively [57]. The diagnostic accuracy of early GC could be increased by the combination of gastric juice ABHD11-AS1, serum, and gastric juice CEA [58].

2.4 Hepatocellular carcinoma (HCC)

Liver cancer is one of the most prevalent cancers around the globe and has the fastest increasing mortality for decades [29]. Hepatocellular carcinoma (HCC) is the most commonly diagnosed type of liver cancer, and early metastasis, high-grade malignancy, as well as lack of proper biomarkers for detection remain the main obstacles to the treatment of HCC [59]. AFP and imaging technology remain main approaches for HCC diagnosis, yet a high missed diagnosis rate of AFP, as well as missed diagnosis when using imaging examinations, still occurs [60, 61].

A number of studies have demonstrated that lncRNAs can be used as biomarkers and prognostic factors in HCC. For example, lnc-PCDH9-13:1 has been reported to be highly expressed in serum and saliva of HCC patients and can be detected in the salivary fluid of HCC patients with early (Stage I and II) and advanced stage (Stage III and IV). Moreover, lnc-PCDH9-13:1 reduced after treatment but significantly increased when treatment failed. The sensitivity and specificity of salivary lnc-PCDH9-13:1 was >80%, much better than the AFP-based HCC diagnosis [62]. ZFAS1 is another lncRNA that is highly upregulated in HCC tissue and serum of HCC patients. The accuracy of ZFAS1 to distinguish HCC patients from healthy individuals was positively associated with serum AFP level [63]. EV-derived LINC00853 showed excellent discriminatory ability in the diagnosis of all-stage HCC (AUC 0.934). Moreover, when diagnosing early-stage HCC, EV-LINC00853 showed higher sensitivity (93.75% versus 9.38%) and specificity (89.77% versus 72.73%) than AFP [64].

In terms of the combination of lncRNA and AFP, it has been indicated that a combination of lncRNA UBE2CP3 and AFP achieved an improved AUC (0.933), compared to lncRNA UBE2CP3 (AUC 0.839) or AFP (AUC 0.912) [65]. Likewise, LINC01793 was highly expressed in the whole blood of HBV-related HCC patients, with AUC value of 0.824, 0.767, and 0.756, respectively, for distinguishing HCC from healthy controls, chronic hepatitis, and liver cirrhosis (LC) patients. The combination of LINC01793 with AFP had a better diagnostic performance than each alone in AFP-negative HCC patients [66]. lncRNA THEMIS2-211 was reported to be highly expressed in HCC tissues, and a panel based on exosomal lncRNA THEMIS2-211 and PRKACA-202 was superior to that of AFP alone in HCC diagnosis. Moreover, exosomal lncRNA THEMIS2-211 alone showed better performance than AFP in early-stage HCC diagnosis [67]. The levels of serum Linc00152 and UCA1 were significantly higher in HCC patients than in patients with benign liver diseases and healthy controls. The combination of serum Linc00152, UCA1, and AFP showed the greatest predictive ability (AUC 0.912, sensitivity 82.9% and specificity 88.2%) [68]. lnc-MyD88 was upregulated in the plasma of hepatitis B virus (HBV)-associated HCC patients. Compared with healthy people or LC patients, lnc-MyD88 showed higher diagnostic value than AFP in HCC patients. The combination of lnc-MyD88 and AFP showed higher AUC, sensitivity, and Youden index than each alone [69].

2.5 Colorectal cancer (CRC)

Colorectal cancer (CRC) is one of the most common human malignancies worldwide and ranks as the second cause of cancer-related mortality in men and women combined (2023. Cancer statistics.). Intensive investigations and advanced therapeutic approaches can help improve CRC patients' survival time. However, due to late diagnosis and treatment, over 50% of CRC patients still face poor prognosis [70]. CEA blood test and colonoscopy remain the most common methods for CRC detection and screening, yet low diagnostic power, high costs, and discomfort produced by invasive procedures make them limited in clinical application [71, 72].

lncRNA CRNDE has several isoforms (CRNDE-a, b, d, e, f, h, and j) and was increased in more than 90% of colorectal adenomas and adenocarcinomas. Compared with healthy individuals, plasma CRNDE-h level was higher in CRC patients (AUC 0.873, sensitivity 87%, and specificity 93%) [73]. Besides, exosomal CRNDE-h had better performance than CEA in CRC patients (AUC 0.892 vs 0.688, sensitivity 70.3% vs 37.16%, and specificity 94.4% vs 88.75%), and the combination of both further enhanced the diagnostic power (AUC 0.913) [74].

In terms of the combination of different lncRNAs, the expression levels of three lncRNAs (TUSC7, RPL34-AS1 and SNHG16) had been reported to be evaluated in the serum samples from patients with CRC. Further analysis showed the combination of these lncRNAs had promising diagnostic potential (AUC 0.793, sensitivity 68.3%, and specificity 86.9%). The combination of these lncRNAs also showed greater diagnostic performance than conventional biomarkers [75]. Another study had reported a diagnostic signature comprising lncRNAs consistently dysregulated in both tumor and serum samples from CRC patients. This signature consisted of four lncRNAs: BANCR (upregulated), LOC148696 (downregulated), LEF1-AS1(downregulated), and LINC00460 (downregulated) [76]. The diagnostic power of this lncRNA panel was greater than that of CEA in stage I-III CRC patients (stage I AUC 0.774 vs 0.588; stage II AUC 0.844 vs 0.695; and stage III AUC 0.949 vs 0.861). Similarly, compared with samples from non-cancerous controls, plasma lncRNAs 91H, PVT-1, and MEG3 were higher in CRC patients. Their combination could discriminate CRC patients from non-cancerous controls (AUC 0.877, sensitivity 82.76% and specificity 78.57%) and showed more sensitivity in the detection of early-stage CRC than the combination of CEA and CA199 [77].

In terms of combination of lncRNAs with conventional biomarkers, serum levels of lncRNA linc01836 in CRC patients were significantly higher than those in healthy donors and benign patients. In addition, serum Linc01836 could effectively differentiate CRC patients from the healthy folks, with the AUC value of 0.809. The combination of serum Linc01836, CEA, and Cyfra21-1 could improve diagnostic sensitivity (92.0%) [78]. Likewise, it was found that the level of lncRNA DANCR was significantly increased in CRC tissue and serum. As for diagnostic efficiency by ROC analysis, the AUC of serum DANCR was higher than that of CEA and CA199 in CRC group vs. colorectal polyp group. The combination of DANCR, CEA, and CA199 showed improved sensitivity and AUC compared with each alone [79].

2.6 Prostate cancer (PCa)

Prostate cancer (PCa) is the most prevalent cancer and ranks as the second-leading cause of cancer-related death among men [29]. PSA is the current serum biomarker for PCa diagnosis. However, despite its high specificity, PSA has low sensitivity and can be dysregulated in other prostate abnormalities such as benign prostate hyperplasia and prostatitis [80]. In addition, the large majority of men with PCa is asymptomatic, which is hard to be early diagnosed based on PSA level and digital rectal examination [81]. Due to the special anatomical location of the prostate tissue, several studies focus on the lncRNAs dysregulated in urine of patients with PCa.

PCA3 is a well-investigated prostate-specific lncRNA that is strongly overexpressed in more than 95 % of primary PCa specimens and metastasis [82]. PCA3 can be detected in urine samples from patients with PCa, and based on this lncRNA, the PROGENSA PCA3 test has been approved by the FDA for the detection of PCa, which is the first and currently only urine-based molecular diagnostic test [83]. PCA3 has been reported to have greater specificity but lower sensitivity for PCa diagnosis than serum PSA [83]. The combination of PCA3 and TMPRSS2–ERG fusion transcripts in urine has been demonstrated to better identify men with high-risk PCa (Gleason score ≥7) and help select biopsies more accurately [84]. Furthermore, the combined test had 93% sensitivity for the detection of high-risk PCa, with increased specificity (33%) than serum PSA alone (17%) [84].

MALAT-1 is highly expressed during PCa progression and is associated with aggressive features of tumor cells. In patients with serum PSA levels of 4–10 ng/ml, urine MALAT-1 levels had a better sensitivity (65.4% vs 61.5%) and specificity (66.7% vs 55.6%) than the percentage of free PSA (%fPSA) [85]. The combination of PCA3 and MALAT-1 had a greater diagnostic performance than current clinical parameters in detecting PCa and clinically significant PCa (Gleason score ≥ 7), with AUC values of 0.828 and 0.832, respectively [86].

Cancer type	LncRNA(s)	Biomarkers combined	Sensitivity (%)	Specificity (%)	AUC	Reference
Non-small-cell lung cancer (NSCLC)	HOTAIR	CEA	N.A.	N.A.	0.841	[33]
	Linc00152	CEA	76	83	0.881	[34]
	GAS5	CEA and CA199	86.76	62.62	0.734	[36]
Gastric cancer (GC)	CTC-497E21.4	CEA and CA199	96.36	42.86	0.896	[52]
	HCP5	CEA and CA199	81	79	0.87	[53]
	FAM49B-AS, GUSBP11, and CTDHUT	CA242 and CA724	93.2	86.6	0.952	[54]
	RP11-731F5.2	CEA and CA199	85.71	61.54	0.84	[55]
	ABHD11-AS1	CEA	N.A.	N.A.	N.A.	[58]
Hepatocellular carcinoma (HCC)	UBE2CP3	AFP	N.A.	N.A.	0.933	[65]
	LINC01793	AFP	73.7	90	0.858	[66]
	Linc00152 and UAC1	AFP	82.9	88.2	0.912	[68]
Colorectal cancer (CRC)	Linc01836	CEA and Cyfra21-1	92	62.3	0.916	[78]
	DANCR	CEA and CA199	87.5	55	0.812	[79]
Prostate cancer (PCa)	PCA3	TMPRSS2–ERG fusion transcripts	92.6	33.4	0.77	[84]
	PCA3	MALAT1	N.A.	N.A.	0.828	[85]

Table 1.
The combination of circulating lncRNAs with other biomarkers and their clinical importance.

The urine levels of FR0348383 have been reported to be an independent predictor of PCa [87, 88]. Moreover, FR0348383 showed greater discriminatory ability than PSA and %fPSA in the detection of PCa in men with serum PSA levels of 4–10 ng/ml (AUC 0.815 vs 0.562 vs 0.599).

2.7 Bladder cancer

Bladder cancer is a common urological malignancy, and urothelial carcinoma (BLCA) accounts for approximately 90% of bladder cancer [89]. Bladder cancer is also commonly divided into two subtypes, muscle-invasive bladder cancer (MIBC) and non-muscle-invasive bladder cancer (NMIBC), for diagnostic, prognostic, and therapeutic purposes [90]. The gold standard for bladder cancer diagnosis remains tissue biopsy via cystoscopy. Other clinical methods commonly used, such as the GICA assay for NMP22 and bladder tumor antigen, have low sensitivity and false results due to hematuria [91–93]. Thus, a more meaningful diagnostic index with balanced sensitivity and specificity is needed to better detect bladder cancer.

The combination of three lncRNAs, MEG3, SNHG16, and MALAT1, was reported to have increased performance for detecting MIBC compared with NMIB tumors, with a C-Index >80%, which is higher than that of urine cytology (C-index 0.620) [94]. Similarly, the urinal exosomal levels of MKLN1-AS, TALAM1, TTN-AS1, and UCA1 were reported to be higher in bladder cancer patients than in controls in training and validation cohort. The combination of these four lncRNAs and NMP22 could classify bladder cancer patients from controls in the training cohort (AUC = 0.850), and a better performance in the validation cohort [95].

In summary, the sensitivity and specificity of any tumor-associated circulating lncRNA alone as a biomarker may remain poor, but the combination of these circulating lncRNAs or with conventional cancer biomarkers can considerably promote the efficiency of cancer detection (**Table 1**). Thus, noninvasive procedures based on these lncRNAs from body liquids need to be improved urgently for routine practice.

3. Circulating lncRNAs as potential biomarkers for cancer prognosis

Apart from the ability of cancer diagnosis, circulating lncRNAs may also have the potential of prediction of cancer prognosis. Although numerous studies have focused on lncRNAs from cancer tissue samples to reflect cancer prognosis [96, 97], lncRNAs from blood-derived samples can also be assessed to predict disease evolution and eventual clinical outcome.

A growing body of literature indicated that aberrant expression of circulating lncRNAs may be functional in human malignancies with clinical implications. For example, exosomal CRNDE-h was reported to be overexpressed in CRC patients compared with healthy individuals and colorectal benign disease patients. Exosomal CRNDE-h level was significantly associated with CRC lymph node metastasis and distant metastasis. Furthermore, the multivariate analysis confirmed exosomal CRNDE-h to be an independent predictor of overall survival in CRC patients [74]. LncRNA FRLnc1 was reported to act as an independent prognostic indicator for the prediction of survival in GC patients [98]. The upregulation of FRLnc1 in GC patient serum exosomes was further confirmed to be significantly associated with clinicopathological properties, including lymph node metastasis and advanced TNM stage [99].

Some studies reported the combination of circulating lncRNAs for prognosis evaluation. A serum 5-lncRNA signature, including lncRNA-LET, PVT1, PANDAR, PTENP1, and linc00963, was reported to be capable of discriminating benign renal tumors from metastatic renal cell carcinoma [100]. Similarly, the circulating 8-lncRNA signature (including ADARB2-AS1, ANRIL, GLIS3-AS1, LINC00472, MEG3, PANDA, PVT1, and UCA1) could distinguish indolent/benign intraductal papillary mucinous neoplasms from aggressive/malignant ones [101] and showed improved accuracy when combined with plasma miRNA data and quantitative radiomic imaging. Another signature comprising nine ferroptosis-related lncRNAs (AL031775.1, AL162586.1, AC034236.2, LINC01004, OCIAD1-AS1, AL136084.3, AP003352.1, Z84484.1, AC022150.2) was recently reported to predict the prognosis of patients with BC [102]. Compared with the low-risk group, the high-risk group had a worse prognosis.

Notably, Lee YR et al. determined the prognostic value of circulating exosomal miRNA-21 and lncRNA-ATB for human HCC prospectively [103]. Both circulating exosomal miRNA-21 and lncRNA-ATB were related to prognostic factors such as TNM stage. Multivariate analysis identified that both higher levels of lncRNA-ATB and miRNA-21 were independent predictors of mortality and disease progression, along with larger tumor size and higher C-reactive protein. Patients with higher circulating levels of exosomal miRNA-21 and lncRNA-ATB showed lower overall survival and progression-free survival. Nevertheless, more prospective studies are warranted to validate prognostic value of circulating lncRNAs.

Overall, circulating lncRNAs have the potential of predicting clinical outcomes of cancer patients. The combination of circulating lncRNAs, miRNAs, conventional cancer biomarkers, as well as quantitative imaging features may further improve the predictive ability. More investigations are needed to truly appreciate the prognosis power of circulating lncRNAs and finally make their clinical practice come true.

4. The detection of circulating lncRNAs

Currently, quantitative reverse transcription PCR (qRT- PCR) assays, microarrays, and next generation sequencing (NGS) are the most frequently used approaches for lncRNA detection (**Table 2**). The accuracy of the detection of circulating lncRNAs varies according to detection methods.

As for qRT- PCR assays, it is quick, highly sensitive, easy to customize with the small amounts of input material, and delivers easily interpretable results. However, the low throughput and the limited number of lncRNAs analyzed are still the obstacles for its routine clinical practice [104, 105]. Besides, when using equal quantities of total RNAs to perform qRT-PCR, it should be noted that the quantity of total RNAs extracted from plasma or serum samples is usually very low and may be hard to detect using spectrophotometers [106]. After adjusting all samples to the sample with the lowest RNA concentration, the output of the subsequent qPCR reaction may reduce. Another choice to perform qRT-PCR is using equal volumes of RNA extracts. This method requires normalization using reference genes when analyzing qPCR data by relative quantification. However, although some reference genes are commonly used (e.g., GAPDH and beta-actin) [107], there is still no consensus on the best reference genes. Moreover, the qRT- PCR assay is not recommended for detection of genes with poor abundance or recurrent variability [107, 108]. Thus, one must test several reference genes and carefully choose the best one(s).

	qRT- PCR	Microarrays	RNA-seq
Throughput	Low	High	High
Accuracy	High	Limited	High
RNA amounts	Small	Small	Large
Others	No consensus on reference genes	Cross-hybridization, or nonspecific hybridization probes	Highly rely on sophisticated bioinformatics analysis
	Not suitable for lncRNAs with poor abundance	Issues associated with probe redundancy and annotation	Hard to apply in liquid biopsy assays
Reference	[104–108]	[109, 110]	[13, 104, 111, 112]

Table 2.
Compendium of different detection methods for circulating lncRNAs.

Another technology commonly used for lncRNAs detection is microarrays. Unlike qRT- PCR, this method can analyze thousands of lncRNAs simultaneously by high-throughput investigation [109]. Nonetheless, microarrays have their own limitations. For example, the accuracy of expression measurements may be limited due to background hybridization, particularly when detecting lncRNAs of low abundance. Moreover, cross-hybridization and limited detection range of probes may occur, as well as issues involving probe redundancy and annotation [110]. Both qRT-PCR and microarrays are limited to the detection of predefined targets.

Recently, with the rapid development of the NGS technologies, the whole transcriptome RNA-seq has become one of the most vital and reliable methods in lncRNA detection [13]. NGS is an efficient method to collect datasets with high accuracy and relatively low costs, yet it still has its disadvantages. For example, the accuracy of data interpretation highly relies on sophisticated bioinformatic analysis, and suitable bioinformatic workflows are required when analyzing the gathered data [104, 111]. Besides, although multiple RNA-seq modifications for lncRNA analysis have been established [112], they are still hard to apply in liquid biopsy assays due to the lack of large amounts of total RNA needed for the input material.

5. Conclusions

Numerous studies have demonstrated that lncRNAs are dysregulated in cancers and can be detected in body fluids; thus, they have the potential to be promising biomarkers of relevant cancers. However, several challenges still present for the clinical application of circulating lncRNAs.

One central concern is to improve their diagnostic performance. Although the combination of multiple lncRNAs and other conventional biomarkers aid to improve the diagnostic potential, it needs to be noted that the "specificity" determined in the reported studies only makes senses when comparing healthy volunteers. The "specificity" may fail to distinguish a certain cancer type from other cancers. Indeed, several lncRNAs have been reported to be potential biomarkers for a variety of cancers [113, 114]. Thus, circulating lncRNAs currently reportedly related to certain cancers may eventually be proved as pan-cancer markers.

The existing methods for lncRNA detection have both advantages and disadvantages at the technic levels. Besides, lncRNAs involved in tumorigenesis and metastasis only account for a small fraction of total lncRNAs discovered to date [11]. Thus, existing detection methods need to be improved to keep pace with the rate at which new lncRNAs are annotated and to functionally screen thousands of lncRNAs at low cost.

In summary, circulating lncRNAs can serve as reliable biomarkers for both cancer diagnosis and prognosis. The development of new technologies to improve their detection will certainly increase the chance of discovering reliable blood-based biomarkers for early and accurate detection of cancers.

Acknowledgements

This work was supported by the National Natural Science Foundation of China (No. 82173345 to Le Qu, No. 81972333 to Cheng Chen), Natural Science Foundation of Jiangsu Province for Distinguished Young Scholars (No. BK20200006 to Le Qu), China National Key Research and Development Program Stem Cell and Translational Research Key Projects (2018YFA0108300 to Le Qu).

Conflict of interest

The authors declare no conflict of interest.

Author details

Le Qu[1*], Cheng Chen[2*], Meijuan Wu[2] and Yulin Zhou[1]

1 Department of Urology, Jinling Hospital, Affiliated Hospital of Medical School, Nanjing University, Nanjing, China

2 Department of Medical Oncology, Jinling Hospital, Affiliated Hospital of Medical School, Nanjing University, Nanjing, China

*Address all correspondence to: septsoul@hotmail.com and chencheng1289@126.com

References

[1] Heery R, Finn SP, Cuffe S, Gray SG. Long non-coding RNAs: Key regulators of epithelial-mesenchymal transition, tumour drug resistance and cancer stem cells. Cancers (Basel). 2017;**9**(4)

[2] Geisler S, Coller J. RNA in unexpected places: Long non-coding RNA functions in diverse cellular contexts. Nature Reviews. Molecular Cell Biology. 2013;**14**(11):699-712

[3] Zhang Q, Wu E, Tang Y, Cai T, Zhang L, Wang J, et al. Deeply mining a universe of peptides encoded by long noncoding RNAs. Molecular & Cellular Proteomics. 2021;**20**:100109

[4] Balas MM, Johnson AM. Exploring the mechanisms behind long noncoding RNAs and cancer. Noncoding RNA Research. 2018;**3**(3):108-117

[5] Li L, Liu B, Wapinski OL, Tsai MC, Qu K, Zhang J, et al. Targeted disruption of Hotair leads to homeotic transformation and gene depression. Cell Reports. 2013;**5**(1):3-12

[6] Mirzaei S, Paskeh MDA, Okina E, Gholami MH, Hushmandi K, Hashemi M, et al. Molecular landscape of LncRNAs in prostate cancer: A focus on pathways and therapeutic targets for intervention. Journal of Experimental & Clinical Cancer Research. 2022;**41**(1):214

[7] Tuck AC, Tollervey D. A transcriptome-wide atlas of RNP composition reveals diverse classes of mRNAs and lncRNAs. Cell. 2013;**154**(5):996-1009

[8] Statello L, Guo CJ, Chen LL, Huarte M. Gene regulation by long non-coding RNAs and its biological functions. Nature Reviews. Molecular Cell Biology. 2021;**22**(2):96-118

[9] Herman AB, Tsitsipatis D, Gorospe M. Integrated lncRNA function upon genomic and epigenomic regulation. Molecular Cell. 2022;**82**(12):2252-2266

[10] Goodall GJ, Wickramasinghe VO. RNA in cancer. Nature Reviews. Cancer. 2021;**21**(1):22-36

[11] Esposito R, Bosch N, Lanzos A, Polidori T, Pulido-Quetglas C, Johnson R. Hacking the cancer genome: Profiling therapeutically actionable long non-coding RNAs using CRISPR-Cas9 screening. Cancer Cell. 2019;**35**(4):545-557

[12] Jin KT, Lu ZB, Lv JQ, Zhang JG. The role of long non-coding RNAs in mediating chemoresistance by modulating autophagy in cancer. RNA Biology. 2020;**17**(12):1727-1740

[13] Anfossi S, Babayan A, Pantel K, Calin GA. Clinical utility of circulating non-coding RNAs – An update. Nature Reviews. Clinical Oncology. 2018;**15**(9):541-563

[14] Arita T, Ichikawa D, Konishi H, Komatsu S, Shiozaki A, Shoda K, et al. Circulating long non-coding RNAs in plasma of patients with gastric cancer. Anticancer Research. 2013;**33**(8):3185-3193

[15] Tsai MC, Manor O, Wan Y, Mosammaparast N, Wang JK, Lan F, et al. Long noncoding RNA as modular scaffold of histone modification complexes. Science. 2010;**329**(5992): 689-693

[16] Gareev I, Ahmad A, Wang J, Beilerli A, Ilyasova T, Sufianov A, et al. Gastric juice non-coding RNAs as

potential biomarkers for gastric cancer. Frontiers in Physiology. 2023;**14**:1179582

[17] Valadi H, Ekstrom K, Bossios A, Sjostrand M, Lee JJ, Lotvall JO. Exosome-mediated transfer of mRNAs and microRNAs is a novel mechanism of genetic exchange between cells. Nature Cell Biology. 2007;**9**(6):654-659

[18] Li Q, Shao Y, Zhang X, Zheng T, Miao M, Qin L, et al. Plasma long noncoding RNA protected by exosomes as a potential stable biomarker for gastric cancer. Tumour Biology. 2015;**36**(3):2007-2012

[19] Sun Z, Yang S, Zhou Q, Wang G, Song J, Li Z, et al. Emerging role of exosome-derived long non-coding RNAs in tumor microenvironment. Molecular Cancer. 2018;**17**(1):82

[20] Sarfi M, Abbastabar M, Khalili E. Long noncoding RNAs biomarker-based cancer assessment. Journal of Cellular Physiology. 2019;**234**(10):16971-16986

[21] Qi P, Zhou XY, Du X. Circulating long non-coding RNAs in cancer: Current status and future perspectives. Molecular Cancer. 2016;**15**(1):39

[22] Wu SG, He ZY, Zhou J, Sun JY, Li FY, Lin Q, et al. Serum levels of CEA and CA15-3 in different molecular subtypes and prognostic value in Chinese breast cancer. Breast. 2014;**23**(1):88-93

[23] Harris L, Fritsche H, Mennel R, Norton L, Ravdin P, Taube S, et al. American Society of Clinical Oncology 2007 update of recommendations for the use of tumor markers in breast cancer. Journal of Clinical Oncology. 2007;**25**(33):5287-5312

[24] Zhang K, Luo Z, Zhang Y, Zhang L, Wu L, Liu L, et al. Circulating lncRNA H19 in plasma as a novel biomarker for breast cancer. Cancer Biomarkers. 2016;**17**(2):187-194

[25] Zhong G, Wang K, Li J, Xiao S, Wei W, Liu J. Determination of serum exosomal H19 as a noninvasive biomarker for breast cancer diagnosis. Oncotargets and Therapy. 2020;**13**:2563-2571

[26] Zhang Y, Zhang K, Luo Z, Liu L, Wu L, Liu J. Circulating long non-coding HOX transcript antisense intergenic ribonucleic acid in plasma as a potential biomarker for diagnosis of breast cancer. Thoracic Cancer. 2016;7(6):627-632

[27] Zidan HE, Karam RA, El-Seifi OS, Abd Elrahman TM. Circulating long non-coding RNA MALAT1 expression as molecular biomarker in Egyptian patients with breast cancer. Cancer Genetics. 2018;**220**:32-37

[28] El-Ashmawy NE, Hussien FZ, El-Feky OA, Hamouda SM, Al-Ashmawy GM. Serum LncRNA-ATB and FAM83H-AS1 as diagnostic/prognostic non-invasive biomarkers for breast cancer. Life Sciences. 2020;**259**:118193

[29] Siegel RL, Miller KD, Wagle NS, Jemal A. Cancer statistics, 2023. CA: A Cancer Journal for Clinicians. 2023;**73**(1):17-48

[30] Coate LE, John T, Tsao MS, Shepherd FA. Molecular predictive and prognostic markers in non-small-cell lung cancer. The Lancet Oncology. 2009;**10**(10):1001-1010

[31] Strauss GM, Kwiatkowski DJ, Harpole DH, Lynch TJ, Skarin AT, Sugarbaker DJ. Molecular and pathologic markers in stage I non-small-cell carcinoma of the lung. Journal of Clinical Oncology. 1995;**13**(5):1265-1279

[32] Tarro G, Perna A, Esposito C. Early diagnosis of lung cancer by detection

of tumor liberated protein. Journal of Cellular Physiology. 2005;**203**(1):1-5

[33] Li N, Wang Y, Liu X, Luo P, Jing W, Zhu M, et al. Identification of circulating long noncoding RNA HOTAIR as a novel biomarker for diagnosis and monitoring of non-small cell lung cancer. Technology in Cancer Research & Treatment. 2017;**16**(6):1060-1066

[34] Li N, Feng XB, Tan Q, Luo P, Jing W, Zhu M, et al. Identification of circulating long noncoding RNA Linc00152 as a novel biomarker for diagnosis and monitoring of non-small-cell lung cancer. Disease Markers. 2017;**2017**:7439698

[35] Liang W, Lv T, Shi X, Liu H, Zhu Q, Zeng J, et al. Circulating long noncoding RNA GAS5 is a novel biomarker for the diagnosis of nonsmall cell lung cancer. Medicine (Baltimore). 2016;**95**(37):e4608

[36] Tan Q, Zuo J, Qiu S, Yu Y, Zhou H, Li N, et al. Identification of circulating long non-coding RNA GAS5 as a potential biomarker for non-small cell lung cancer diagnosisnon-small cell lung cancer, long non-coding RNA, plasma, GAS5, biomarker. International Journal of Oncology. 2017;**50**(5):1729-1738

[37] Yuan S, Xiang Y, Guo X, Zhang Y, Li C, Xie W, et al. Circulating long noncoding RNAs act as diagnostic biomarkers in non-small cell lung cancer. Frontiers in Oncology. 2020;**10**:537120

[38] Min L, Zhu T, Lv B, An T, Zhang Q, Shang Y, et al. Exosomal LncRNA RP5-977B1 as a novel minimally invasive biomarker for diagnosis and prognosis in non-small cell lung cancer. International Journal of Clinical Oncology. 2022;**27**(6):1013-1024

[39] He C, Huang D, Yang F, Huang D, Cao Y, Peng J, et al. High expression of lncRNA HEIH is helpful in the diagnosis of non-small cell lung cancer and predicts poor prognosis. Cancer Management and Research. 2022;**14**:503-514

[40] Thrift AP, Wenker TN, El-Serag HB. Global burden of gastric cancer: Epidemiological trends, risk factors, screening and prevention. Nature Reviews. Clinical Oncology. 2023;**20**(5):338-349

[41] Jemal A, Siegel R, Ward E, Murray T, Xu J, Thun MJ. Cancer statistics, 2007. CA: A Cancer Journal for Clinicians. 2007;**57**(1):43-66

[42] Kang F, Jiang F, Ouyang L, Wu S, Fu C, Liu Y, et al. Potential biological roles of exosomal long non-coding RNAs in gastrointestinal cancer. Frontiers in Cell and Development Biology. 2022;**10**:886191

[43] Zhao R, Zhang Y, Zhang X, Yang Y, Zheng X, Li X, et al. Exosomal long noncoding RNA HOTTIP as potential novel diagnostic and prognostic biomarker test for gastric cancer. Molecular Cancer. 2018;**17**(1):68

[44] Xian HP, Zhuo ZL, Sun YJ, Liang B, Zhao XT. Circulating long non-coding RNAs HULC and ZNFX1-AS1 are potential biomarkers in patients with gastric cancer. Oncology Letters. 2018;**16**(4):4689-4698

[45] Jin C, Shi W, Wang F, Shen X, Qi J, Cong H, et al. Long non-coding RNA HULC as a novel serum biomarker for diagnosis and prognosis prediction of gastric cancer. Oncotarget. 2016;**7**(32):51763-51772

[46] Feng W, Zong W, Li Y, Shen X, Cui X, Ju S. Abnormally expressed long noncoding RNA B3GALT5-AS1 may serve as a biomarker for the diagnostic

and prognostic of gastric cancer. Journal of Cellular Biochemistry. 2020;**121**(1):557-565

[47] Liu W, Li Y, Zhang Y, Shen X, Su Z, Chen L, et al. Circulatinglong non-coding RNA FEZF1-AS1 and AFAP1-AS1 serve as potential diagnostic biomarkers for gastric cancer. Pathology, Research and Practice. 2020;**216**(1):152757

[48] Cai C, Zhang H, Zhu Y, Zheng P, Xu Y, Sun J, et al. Serum exosomal long noncoding RNA pcsk2-2:1 as a potential novel diagnostic biomarker for gastric cancer. Oncotargets and Therapy. 2019;**12**:10035-10041

[49] Li S, Zhang M, Zhang H, Hu K, Cai C, Wang J, et al. Exosomal long noncoding RNA lnc-GNAQ-6:1 may serve as a diagnostic marker for gastric cancer. Clinica Chimica Acta. 2020;**501**:252-257

[50] Zheng P, Zhang H, Gao H, Sun J, Li J, Zhang X, et al. Plasma exosomal long noncoding RNA lnc-SLC2A12-10:1 as a novel diagnostic biomarker for gastric cancer. Oncotargets and Therapy. 2020;**13**:4009-4018

[51] Guo X, Lv X, Ru Y, Zhou F, Wang N, Xi H, et al. Circulating exosomal gastric cancer-associated long noncoding RNA1 as a biomarker for early detection and monitoring progression of gastric cancer: A multiphase study. JAMA Surgery. 2020;**155**(7):572-579

[52] Zong W, Feng W, Jiang Y, Ju S, Cui M, Jing R. Evaluating the diagnostic and prognostic value of serum long non-coding RNA CTC-497E21.4 in gastric cancer. Clinical Chemistry and Laboratory Medicine. 2019;**57**(7):1063-1072

[53] Qin S, Yang L, Kong S, Xu Y, Liang B, Ju S. LncRNA HCP5: A potential biomarker for diagnosing gastric cancer. Frontiers in Oncology. 2021;**11**:684531

[54] Zheng R, Liang J, Lu J, Li S, Zhang G, Wang X, et al. Genome-wide long non-coding RNAs identified a panel of novel plasma biomarkers for gastric cancer diagnosis. Gastric Cancer. 2019;**22**(4):731-741

[55] Jing R, Liu S, Jiang Y, Zong W, Ju S, Cui M. Determination of serum RP11-731F5.2 as a noninvasive biomarker for gastric cancer diagnosis and prognosis. Pathology, Research and Practice. 2020;**216**(12):153261

[56] Sung J, Kim N, Lee J, Hwang YJ, Kim HW, Chung JW, et al. Associations among gastric juice pH, atrophic gastritis, intestinal metaplasia and *Helicobacter pylori* infection. Gut Liver. 2018;**12**(2):158-164

[57] Shao Y, Ye M, Jiang X, Sun W, Ding X, Liu Z, et al. Gastric juice long noncoding RNA used as a tumor marker for screening gastric cancer. Cancer. 2014;**120**(21):3320-3328

[58] Yang Y, Shao Y, Zhu M, Li Q, Yang F, Lu X, et al. Using gastric juice lncRNA-ABHD11-AS1 as a novel type of biomarker in the screening of gastric cancer. Tumour Biology. 2016;**37**(1):1183-1188

[59] Bruix J, Sherman M. American association for the study of liver D. management of hepatocellular carcinoma: An update. Hepatology. 2011;**53**(3):1020-1022

[60] Huang L, Mo Z, Hu Z, Zhang L, Qin S, Qin X, et al. Diagnostic value of fibrinogen to prealbumin ratio and gamma-glutamyl transpeptidase to platelet ratio in the progression of AFP-negative hepatocellular carcinoma. Cancer Cell International. 2020;**20**:77

[61] Song P, Cai Y, Tang H, Li C, Huang J. The clinical management of

hepatocellular carcinoma worldwide: A concise review and comparison of current guidelines from 2001 to 2017. Bioscience Trends. 2017;**11**(4):389-398

[62] Xie Z, Zhou F, Yang Y, Li L, Lei Y, Lin X, et al. Lnc-PCDH9-13:1 Is a hypersensitive and specific biomarker for early hepatocellular carcinoma. eBioMedicine. 2018;**33**:57-67

[63] Luo P, Liang C, Zhang X, Liu X, Wang Y, Wu M, et al. Identification of long non-coding RNA ZFAS1 as a novel biomarker for diagnosis of HCC. Bioscience Reports. 2018;**38**(4):BSR20171359

[64] Kim SS, Baek GO, Ahn HR, Sung S, Seo CW, Cho HJ, et al. Serum small extracellular vesicle-derived LINC00853 as a novel diagnostic marker for early hepatocellular carcinoma. Molecular Oncology. 2020;**14**(10):2646-2659

[65] Cao SW, Huang JL, Chen J, Hu YW, Hu XM, Ren TY, et al. Long non-coding RNA UBE2CP3 promotes tumor metastasis by inducing epithelial-mesenchymal transition in hepatocellular carcinoma. Oncotarget. 2017;**8**(39):65370-65385

[66] Mo C, Wu J, Sui J, Deng Y, Li M, Cao Z, et al. Long non-coding RNA LINC01793 as a potential diagnostic biomarker of hepatitis B virus-related hepatocellular carcinoma. Clinical Biochemistry. 2022;**108**:56-62

[67] Yao J, Hua X, Shi J, Hu X, Lui K, He K, et al. LncRNA THEMIS2-211, a tumor-originated circulating exosomal biomarker, promotes the growth and metastasis of hepatocellular carcinoma by functioning as a competing endogenous RNA. The FASEB Journal. 2022;**36**(4):e22238

[68] Huang J, Zheng Y, Xiao X, Liu C, Lin J, Zheng S, et al. A circulating long noncoding RNA panel serves as a diagnostic marker for hepatocellular carcinoma. Disease Markers. 2020;**2020**:5417598

[69] Wang Z, Gao P, Sun W, Rehman AU, Jiang J, Xu S, et al. Long noncoding RNA MyD88 functions as a promising diagnostic biomarker in hepatocellular carcinoma. Front Endocrinol (Lausanne). 2023;**14**:938102

[70] Yu J, Feng Q, Wong SH, Zhang D, Liang QY, Qin Y, et al. Metagenomic analysis of faecal microbiome as a tool towards targeted non-invasive biomarkers for colorectal cancer. Gut. 2017;**66**(1):70-78

[71] Liang Y, Wang W, Fang C, Raj SS, Hu WM, Li QW, et al. Clinical significance and diagnostic value of serum CEA, CA19-9 and CA72-4 in patients with gastric cancer. Oncotarget. 2016;**7**(31):49565-49573

[72] Rahm AK, Cragun D, Hunter JE, Epstein MM, Lowery J, Lu CY, et al. Implementing universal Lynch syndrome screening (IMPULSS): Protocol for a multi-site study to identify strategies to implement, adapt, and sustain genomic medicine programs in different organizational contexts. BMC Health Services Research. 2018;**18**(1):824

[73] Graham LD, Pedersen SK, Brown GS, Ho T, Kassir Z, Moynihan AT, et al. Colorectal neoplasia differentially expressed (CRNDE), a novel gene with elevated expression in colorectal adenomas and adenocarcinomas. Genes & Cancer. 2011;**2**(8):829-840

[74] Liu T, Zhang X, Gao S, Jing F, Yang Y, Du L, et al. Exosomal long noncoding RNA CRNDE-h as a novel serum-based biomarker for diagnosis and prognosis of colorectal cancer. Oncotarget. 2016;**7**(51):85551-85563

[75] Wang C, Yu J, Han Y, Li L, Li J, Li T, et al. Long non-coding RNAs LOC285194, RP11-462C24.1 and Nbla12061 in serum provide a new approach for distinguishing patients with colorectal cancer from healthy controls. Oncotarget. 2016;**7**(43):70769-70778

[76] Wang R, Du L, Yang X, Jiang X, Duan W, Yan S, et al. Identification of long noncoding RNAs as potential novel diagnosis and prognosis biomarkers in colorectal cancer. Journal of Cancer Research and Clinical Oncology. 2016;**142**(11):2291-2301

[77] Liu H, Ye D, Chen A, Tan D, Zhang W, Jiang W, et al. A pilot study of new promising non-coding RNA diagnostic biomarkers for early-stage colorectal cancers. Clinical Chemistry and Laboratory Medicine. 2019;**57**(7):1073-1083

[78] Shen L, Zong W, Feng W, Chen E, Ma S, Yuan J, et al. Upregulated Linc01836 in serum promisingly serving as a diagnostic and prognostic biomarker for colorectal cancer. Frontiers in Pharmacology. 2022;**13**:840391

[79] Shen X, Xue Y, Cong H, Wang X, Fan Z, Cui X, et al. Circulating lncRNA DANCR as a potential auxillary biomarker for the diagnosis and prognostic prediction of colorectal cancer. Bioscience Reports. 2020;**40**(3):BSR20191481

[80] Lilja H, Ulmert D, Vickers AJ. Prostate-specific antigen and prostate cancer: Prediction, detection and monitoring. Nature Reviews. Cancer. 2008;**8**(4):268-278

[81] Armstrong BK, Barry MJ, Frydenberg M, Gardiner RA, Haines I, Carter SM. PSA testing for men at average risk of prostate cancer. Public Health Research and Practice. 2017;**27**(3):2731721

[82] Hessels D, Klein Gunnewiek JM, van Oort I, Karthaus HF, van Leenders GJ, van Balken B, et al. DD3(PCA3)-based molecular urine analysis for the diagnosis of prostate cancer. European Urology. 2003;**44**(1):8-15. discussion -6

[83] Lee GL, Dobi A, Srivastava S. Prostate cancer: Diagnostic performance of the PCA3 urine test. Nature Reviews. Urology. 2011;**8**(3):123-124

[84] Sanda MG, Feng Z, Howard DH, Tomlins SA, Sokoll LJ, Chan DW, et al. Association between combined TMPRSS2:ERG and PCA3 RNA urinary testing and detection of aggressive prostate cancer. JAMA Oncology. 2017;**3**(8):1085-1093

[85] Ren S, Wang F, Shen J, Sun Y, Xu W, Lu J, et al. Long non-coding RNA metastasis associated in lung adenocarcinoma transcript 1 derived miniRNA as a novel plasma-based biomarker for diagnosing prostate cancer. European Journal of Cancer. 2013;**49**(13):2949-2959

[86] Li Y, Ji J, Lyu J, Jin X, He X, Mo S, et al. A novel urine exosomal lncRNA assay to improve the detection of prostate cancer at initial biopsy: A retrospective multicenter diagnostic feasibility study. Cancers (Basel). 2021;**13**(16):4075

[87] Ren S, Peng Z, Mao JH, Yu Y, Yin C, Gao X, et al. RNA-seq analysis of prostate cancer in the Chinese population identifies recurrent gene fusions, cancer-associated long noncoding RNAs and aberrant alternative splicings. Cell Research. 2012;**22**(5):806-821

[88] Zhang W, Ren SC, Shi XL, Liu YW, Zhu YS, Jing TL, et al. A novel urinary long non-coding RNA transcript improves diagnostic accuracy in patients undergoing prostate biopsy. The Prostate. 2015;**75**(6):653-661

[89] Sung H, Ferlay J, Siegel RL, Laversanne M, Soerjomataram I, Jemal A, et al. Global cancer statistics 2020: GLOBOCAN estimates of incidence and mortality worldwide for 36 cancers in 185 countries. CA: A Cancer Journal for Clinicians. 2021;**71**(3):209-249

[90] Sanli O, Dobruch J, Knowles MA, Burger M, Alemozaffar M, Nielsen ME, et al. Bladder cancer. Natural Review Disease in Primers. 2017;**3**:17022

[91] Zibelman M, Asghar AM, Parker DC, O'Neill J, Wei S, Greenberg RE, et al. Cystoscopy and systematic bladder tissue sampling in predicting pT0 bladder cancer: A prospective trial. The Journal of Urology. 2021;**205**(6):1605-1611

[92] Hwang EC, Choi HS, Jung SI, Kwon DD, Park K, Ryu SB. Use of the NMP22 BladderChek test in the diagnosis and follow-up of urothelial cancer: A cross-sectional study. Urology. 2011;**77**(1):154-159

[93] Sarosdy MF, deVere White RW, Soloway MS, Sheinfeld J, Hudson MA, Schellhammer PF, et al. Results of a multicenter trial using the BTA test to monitor for and diagnose recurrent bladder cancer. The Journal of Urology. 1995;**154**(2 Pt 1):379-383. discussion 83-4

[94] Duan W, Du L, Jiang X, Wang R, Yan S, Xie Y, et al. Identification of a serum circulating lncRNA panel for the diagnosis and recurrence prediction of bladder cancer. Oncotarget. 2016;**7**(48):78850-78858

[95] Bian B, Li L, Ke X, Chen H, Liu Y, Zheng N, et al. Urinary exosomal long non-coding RNAs as noninvasive biomarkers for diagnosis of bladder cancer by RNA sequencing. Frontiers in Oncology. 2022;**12**:976329

[96] Chen S, Shen X. Long noncoding RNAs: Functions and mechanisms in colon cancer. Molecular Cancer. 2020;**19**(1):167

[97] Liu SJ, Dang HX, Lim DA, Feng FY, Maher CA. Long noncoding RNAs in cancer metastasis. Nature Reviews. Cancer. 2021;**21**(7):446-460

[98] Chong DQ, Shan JL, Yang CS, Wang R, Du ZM. Clinical prognostic value of A FOXM1 related long non-coding RNA expression in gastric cancer. European Review for Medical and Pharmacological Sciences. 2018;**22**(2):417-421

[99] Zhang Y, Chen L, Ye X, Wu Z, Zhang Z, Sun B, et al. Expression and mechanism of exosome-mediated A FOXM1 related long noncoding RNA in gastric cancer. Journal of Nanobiotechnology. 2021;**19**(1):133

[100] Wu Y, Wang YQ, Weng WW, Zhang QY, Yang XQ, Gan HL, et al. A serum-circulating long noncoding RNA signature can discriminate between patients with clear cell renal cell carcinoma and healthy controls. Oncogene. 2016;**5**(2):e192

[101] Permuth JB, Chen DT, Yoder SJ, Li J, Smith AT, Choi JW, et al. Lincing circulating long non-coding RNAs to the diagnosis and malignant prediction of intraductal papillary mucinous neoplasms of the pancreas. Scientific Reports. 2017;7(1):10484

[102] Chen M, Nie Z, Li Y, Gao Y, Wen X, Cao H, et al. A new ferroptosis-related lncRNA signature predicts the prognosis of bladder cancer patients. Frontiers in Cell and Development Biology. 2021;**9**:699804

[103] Lee YR, Kim G, Tak WY, Jang SY, Kweon YO, Park JG, et al. Circulating

exosomal noncoding RNAs as prognostic biomarkers in human hepatocellular carcinoma. International Journal of Cancer. 2019;**144**(6):1444-1452

[104] Fernandez-Mercado M, Manterola L, Larrea E, Goicoechea I, Arestin M, Armesto M, et al. The circulating transcriptome as a source of non-invasive cancer biomarkers: Concepts and controversies of non-coding and coding RNA in body fluids. Journal of Cellular and Molecular Medicine. 2015;**19**(10):2307-2323

[105] Benes V, Castoldi M. Expression profiling of microRNA using real-time quantitative PCR, how to use it and what is available. Methods. 2010;**50**(4):244-249

[106] Badowski C, He B, Garmire LX. Blood-derived lncRNAs as biomarkers for cancer diagnosis: The good, the bad and the beauty. NPJ Precision Oncology. 2022;**6**(1):40

[107] Dong L, Qi P, Xu MD, Ni SJ, Huang D, Xu QH, et al. Circulating CUDR, LSINCT-5 and PTENP1 long noncoding RNAs in sera distinguish patients with gastric cancer from healthy controls. International Journal of Cancer. 2015;**137**(5):1128-1135

[108] Gresner P, Gromadzinska J, Wasowicz W. Reference genes for gene expression studies on non-small cell lung cancer. Acta Biochimica Polonica. 2009;**56**(2):307-316

[109] Shi J, Li X, Zhang F, Zhang C, Guan Q, Cao X, et al. Circulating lncRNAs associated with occurrence of colorectal cancer progression. American Journal of Cancer Research. 2015;**5**(7):2258-2265

[110] Zhao S, Fung-Leung WP, Bittner A, Ngo K, Liu X. Comparison of RNA-Seq and microarray in transcriptome profiling of activated T cells. PLoS One. 2014;**9**(1):e78644

[111] Beermann J, Piccoli MT, Viereck J, Thum T. Non-coding RNAs in development and disease: Background, mechanisms, and therapeutic approaches. Physiological Reviews. 2016;**96**(4):1297-1325

[112] Kashi K, Henderson L, Bonetti A, Carninci P. Discovery and functional analysis of lncRNAs: Methodologies to investigate an uncharacterized transcriptome. Biochimica et Biophysica Acta. 2016;**1859**(1):3-15

[113] Hussen BM, Azimi T, Abak A, Hidayat HJ, Taheri M, Ghafouri-Fard S. Role of lncRNA BANCR in human cancers: An updated review. Frontiers in Cell and Development Biology. 2021;**9**:689992

[114] Shen S, Wang Y, Zhang Y, Dong Z, Xing J. Long non-coding RNA small nucleolar RNA host gene 14, a promising biomarker and therapeutic target in malignancy. Frontiers in Cell and Development Biology. 2021;**9**:746714

Section 2

Noncoding RNA in Development

Chapter 5

Shedding Light on the Dark Matter, Noncoding RNAs in Human Reproduction

Rana Alhamdan

Abstract

Advances in human transcriptome have unveiled the crucial regulatory role of noncoding RNA (ncRNA) in most biological processes, including reproduction. Recent studies have elucidated some of the questions, highlighting the regulatory function of specific ncRNAs on concrete reproductive mechanisms. ncRNAs have been shown to be crucial for the maintenance of spermatogenesis, primordial germ cells (PGCs) survivals, folliculogenesis, oocyte maturation, and corpus luteum function. In addition, due to their unique expression and critical functions, they have been demonstrated to be associated with aspects of infertility such as premature ovarian failure (POF), recurrent implantation failure (RIF), polycystic ovarian syndrome (POCS), varicocele, sperm abnormalities, and testicular cancer. This chapter will discuss the current knowledge of the role of ncRNAs in spermatogenesis, and oogenesis and their potential utilization as a noninvasive diagnostic marker for reproductive disorders.

Keywords: ncRNA, folliculogenesis, spermatogenesis, reproductive disorders, miRNAs, lncRNAs, piRNAs

1. Introduction

Human reproduction is made of intricate but interconnected cellular and molecular processes through which a viable healthy offspring is born. Recent studies have suggested that the disruptive status of reproductive health has led to infertility-related problems. The World Health Organization (WHO) has recently prognosticated infertility to might be considered as a major health concern after cancer and cardiovascular disease [1]. Environmental and physiological disruptors render the reproductive process particularly susceptible to epigenetic alterations and modifications. Epigenetic changes have been indicated to play an important role in gametogenesis and reproductive-related disorders. Advanced technologies have proved ncRNA crucial in controlling reproductive functions. Mounting evidence indicates that ncRNAs in gametes are sensitive to environmentally derived epigenetic changes and can moderate the inheritance of paternally acquired disorders attributes [2]. The different types of ncRNAs, dynamic expression, and the complexity of the regulatory pathways have been described, such as the dark matter of the universe.

However, recent studies have shed some light on modulatory functions of specific ncRNAs on concrete reproductive mechanisms. ncRNAs have been shown to participate in multiple levels of gene expression moderation in oogenesis and spermatogenesis such as gene imprinting, meiosis, cell proliferation, and differentiation [3]. ncRNAs have been reported to be crucial for the maintenance of spermatogenesis, primordial germ cells (PGCs) survivals, folliculogenesis, oocyte maturation, and corpus luteum functions [4]. In addition, due to their unique expression and critical functions, they have been demonstrated to be associate with infertility such as premature ovarian failure (POF), recurrent implantation failure (RIF), polycystic ovarian syndrome (POCS), varicocele, sperm abnormalities, and testicular cancer. However, there are many challenges to overcome to elucidate the actual effect of ncRNA aberrancy on reproductive biology and disease. This chapter aims to summarize the current knowledge of ncRNA-related reproductive processes. Here, we will also review recent discoveries linking ncRNAs to infertility-related disorders. An enhanced understanding of the role of ncRNAs can further advance the potential utilization of ncRNAs as a noninvasive diagnostic marker and therapeutic target for reproductive disorders, which will be covered in this chapter as well.

1.1 Overview of ncRNA

A decades-long discussion has been escalated with the surprising, challenging discovery of ncRNAs. Genome-wide sequencing technology and high throughput sequencing revealed that the majority (~93%) of the human genome is actually transcribed into RNA. However, only 2% of this RNA encodes protein [5–7]. Those RNA sequences that were previously described as gene transcriptional "noise" or junk DNA are now believed to play a very important role in human normal cells physiology and pathological processes [7]. The rapid progress and accessibility to the technology have allowed a deeper understanding of transcription on a wider scale and in different conditions. Next generation sequencing (NGS), in particular, offered the possibility to identify and classify noncoding RNAs [8, 9]. ncRNAs have been ranked based on their function, biogenesis, and size in two broad subfamilies: small ncRNA (sncRNA) with size <200 nucleotides (nts) long and include housekeeping ncRNAs, comprising small nuclear RNAs (snRNAs), small nucleolar RNAs (snoRNAs), and other regulatory sncRNAs such as microRNA (miRNA), piwi-interacting RNAs (piRNAs), small endogenous interfering RNAs (endo-siRNAs), and transfer RNA (tRNA). Long ncRNA defines the second group with a size >200 nts in length [10–12].

1.2 Definition, biogenesis, and functions of noncoding RNA

It can be anticipated that the most well-characterized sncRNAs are miRNAs, endo-siRNA, and piRNA. All the three types act *via* an RNA interference (RNAi) key gene regulatory pathway [13]. Fundamental to this pathway is the generation of small noncoding transcripts to assemble the RNA-induced silencing complex (RISC) by Argonaute proteins to cleave target mRNA. Two subfamilies of Argonaute are known as the AGO and PIWI families. AGO protein binds to both miRNA and siRNA, whereas the PIWI subfamily is specific for piRNAs [13].

miRNAs are short (21–22 nts) single-stranded ncRNAs that regulate gene expression posttranscriptionally *via* promoting mRNA translation silencing or degradation [4, 11]. miRNAs are initially derived from nuclear long primary transcripts that are then cleaved by nuclear RNA polymerase III protein complex containing Dorsha and

DiGeotge critical region 8 (DGCR8) into 70 nt hairpin precursor miRNAs (pre-miRNAs). These small transcripts subsequently then leave the nucleus to the cytoplasm and are processed further by Dicer, another member of the nuclear protein complex, to form a mature double-stranded miRNA of 22 nt. The duplexes are then incorporated into a ribonuclear vesicle to assemble miRNA-RISC (miRISC) by Ago subfamily protein, where the complementary strand is degraded, and the mature miRNA form is retained. Mature miRNA can read and bind to the 3'end of their mRNA target by the first eight residues on the 5'end known as the "seed sequence," leading to translation suppression and/or degradation of the target mRNA [14–16]. miRNAs embrace 1–5% of the genome and can typically modulate the expression of myriads of genes. Interestingly, it has also been reported that a single gene can be the target of multiple miRNAs, which, in fact, further complicates their regulatory pathways. Expression alteration of a single miRNA can lead to a profound impact on various cellular functions [11, 15, 17].

Similar to miRNAs, endogenous siRNA plays essential roles to posttranscriptionally regulating gene expression. It functions *via* binding to the 3' UTR site of the target mRNA; however, unlike miRNA, it requires a full complementary sequence in order to silence mRNA translation or promote its decay [18]. siRNA exhibits the same size as miRNA, and both are processed from a hairpin double-strand precursors by Dicer-dependent mechanism [19]. piRNA are marginally longer (24-31 nt) than siRNA and miRNA. In contrast to miRNA and siRNA, piRNAs are processed from long single-stranded precursors by Dicer-independent mechanism. These transcripts are thought to be exported to the cytoplasm and subsequently processed into a primary piRNA [19–21]. In the cytoplasm, the secondary piRNA is generated when the primary piRNA loads onto the PIWI proteins and the homologous sequence. The later will preferentially bind to Piwi family proteins belonging to the Aubergine (Aub) superfamily, exciting the cleavage and generation of more piRNAs *via* the processing of antisense RNAs sequences [20]. Rounds of cleavage repeats produce several piRNAs in a process called a ping-pong cycle. The produced transcripts are then transported back to the nucleus by PIWI proteins to silence target genes. The Aub-piRNA, however, acts as a trigger to the ping-pong amplification cycle, by which, transposons are silenced and piRNAs are amplified.

The biogenesis of piRNA is elusive still and most of these data were from Drosophila which is believed similar to humans. piRNAs were ultimately found in germ cells and were reported to play a key role in germline specification, maintenance, and gametogenesis [11, 21]. A great deal of data is indicating a crucial regulatory function of piRNA in somatic cells *via* mRNA turnover, DNA rearrangement, transposon expression and silencing, translational modulation, and epigenetic programming. The dysregulation of piRNA has been shown to contribute to epigenetic modification-related disease stat [13, 20]. Recent studies demonstrated a differential expression of the PIWI-related proteins in the oocytes and surrounding granulosa cells (GC) in cases with diminished ovarian reserve (DOR). It's interesting to note that PIWI mutant female mice seem to be unaffected [22], while the mutant male mice experience spermatogonia degeneration and altered spermatogenesis [11, 23]. Recent studies have indicated that the mouse genome encodes three PIWI family proteins named MIWI, MILI, and MIWI2. MIWI and MILI act as an RNA-slicing component crucial for retrotransposon inhibition and piRNA biogenesis. The loss of function of any of these three genes leads to spermatogenesis defects [24]. Tudor domain-containing (TDRD) protein has also been reported as essential for piRNA biosynthesis and repression of retrotransposon when complex with PIWI proteins [24, 25].

The biosynthesis of lncRNA in mammals has been preliminarily elucidated. It was indicated that many lncRNAs are transcribed by RNA polymerase II, followed by posttranscriptional 5′-cap addition [1, 26]. It's fascinating to note that lnRNAs are generated through pathways similar to those for mRNAs, with comparable cleavage signals, intron/exon lengths, and histone-modification [27, 28]. According to study by Ma et al., in 2013, lnRNAs have been found to possess similar transcriptional activity [29]. However, other lncRNAs are processed from precursors into small transcripts of unknown function [28]. LncRNA can be classified based on its location within the genome into; 1. Long intergenic (LincRNA), which does not overlap protein-coding genes, 2. Sense intronic transcripts located within introns of a coding gene, however, do not overlap with any exons, 3. Sense intersecting transcripts from introns of protein-coding locus, 4. Antisense RNA, overlapping any exon of a protein-coding gene on the opposite strand, and 5. Spliced transcripts contain no open reading frame (ORF) and do not match any of the other categories [27–29]. The nucleus is the location of most lncRNAs, where they act as a scaffold to assist in chromatin structure modification or alternative splicing [11, 30]. However, it has recently been suggested that specific lncRNAs reside and function in the cytoplasm. They act to promote or inhibit mRNA decay, miRNA sponges, and regulate translation [11, 31, 32]. LncRNAs exhibit poor interspecies conservation, leading to uncertainty about their functionality. Possibly due to their low expression, leading to the proposed view of them being transcriptional noise [29]. Despite this, lncRNAs have been reported to interact with RNA, DNA, and proteins and can significantly impact biological processes such as protein localization, transcription, translation, splicing, imprinting, and reprogramming [11, 29]. It has been suggested to play an important role in infertility, reproduction-related disorders, and cancer disease progression [1, 29, 33]. The role of ncRNAs in regulating ovarian follicular development, spermatogenesis, as well as reproductive disease will be the next focus of this chapter.

A simple comparison of all three ncRNAs is included in **Table 1**.

	miRNA	piRNA	lncRNA
Length (nt)	~22	23–31	~200–800
Expression	All tissues	Germ cell line mainly	All tissues
Precursors	Transcripts with hairpins	Single strand transcripts	transcribed from intergenic regions and from introns of protein-coding genes
Drosha dependent	Yes	No	No
Dicer dependent	Yes	No	No
Argonaute proteins involved	AGO subfamily	PIWI subfamily	No
Function	Translational inhibition, mRNA degradation and storage	Transposable elements cleavage, Translational regulation, and DNA methylation	Cis or trans, splicing, translational, transcriptional and post-transcriptional regulation
Species specificity	Highly conserved across different species	Lack of conservation across the species	Less degree of sequence conservation across the species

Table 1.
Comparison of three the ncRNAs [32, 34–36].

2. ncRNA in reproduction

Germ cell replication and differentiation is a critical step during gametogenesis. Successful transcriptions of specific genes are fundamental for this process. However, the occurrence of any errors during this process can be the principal cause of infertility. [12, 18]. Studies indicate that ncRNAs are relevant from very early stages of germ cell development. They have been demonstrated to play a critical role in regulating transcriptional, posttranscriptional gene expression, and epigenetic-related mechanisms during spermatogenesis and oogenesis [18]. ncRNAs can intriguingly perform dimorphic distinct functions between males and females and have been suggested as a biomarker for oocyte and sperm quality. Here, we are focusing on miRNAs, piRNAs, and lncRNAs, as the most germline related classes of ncRNAs.

2.1 Roles of noncoding RNAs in sperm maturation and functions

Spermatogenesis is the process by which spermatogonial stem cells (SSCs) actively proliferate and differentiate into haploid male gametes. During sperm development, posttranscriptional regulation is particularly crucial at the later stages of their growth. ncRNAs have been shown to be critically important during this complex process at the transcriptional and posttranscriptional level. The process of spermatogenesis can be divided into three phases based on the relevance of ncRNAs: germ cell proliferation and spermatogonia formation, two rounds of meiosis to produce haploid round spermatid, and spermiogenesis (sperm maturation (**Figure 1**) [1].

Hundreds of ncRNAs have recently been discovered throughout spermatogenesis. Of that, 7% of miRNAs were found to be selectively expressed in mouse and human

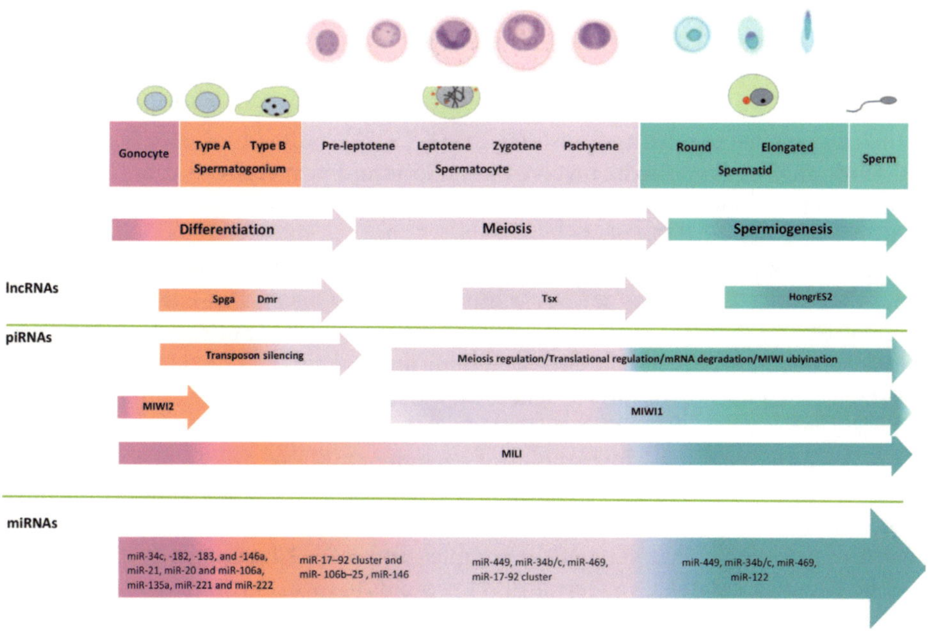

Figure 1.
Characterizations expression of miRNAs, piRNA, and lncRNAs during the different stages of spermatogenesis [12, 18, 37].

immature and mature spermatozoa, as well as in the testis [1, 33, 38, 39], suggesting a crucial role of miRNA at almost every stage of sperm cell development. In mice germ cells, the inhibition of miRNA biogenesis components and therefore miRNA led to male infertility. A recent study by [40] further supported these findings. This study and two others indicated that miRNA processing key factors, in particular Dorsha and Dicer, depletion in Sertoli cells severely impaired testicular function, led to male infertility, oligoteratozoospermia, or azoospermia [40–42]. Moreover, the roles of several miRNAs have been described and pinpointed at different stages of spermatogenesis.

During the early stages (first phase of spermatogenesis), numerous miRNAs have been found to be highly expressed in mouse SSCs, such as miR-34c, miR-21, miR-146a miR-182, miR-183, and clusters of miR-17-92 and miRNA 290-295 [1, 14, 18]. The regulatory role of most of these miRNAs remains to be elucidated. miR-34c was particularly reported to promote SSCs differentiation and may affect their apoptosis [18, 43]. miR-21, miR-20, and miR-106 were found to play an important role in SSC self-renewal and differentiation via regulating homeostasis [18, 44]. Additionally, miR-10b has also been shown to act *via* targeting KLF4 to control SSC self-renewal, whereas miR-202-3p is involved in initiating spermatogonial meiosis [45, 46]. Niu and colleagues demonstrated a regulatory role of miR-204 in SSC proliferation *via* targeting Sirt1 [47]. Other miRNAs, such as miR-221/222, miR-146 act to maintain SSCs in an undifferentiated state [14, 18, 48].

The effect of many miRNAs during the later stages of spermatogenesis is well documented, at which they act *via* different targets and exhibit a regulatory role in meiosis and spermatocyte differentiation. miR-10a is a good example, demonstrating a critical modulatory role during human male germ cell meiosis *via* down-regulating RAD51. The overexpression of this miRNA may lead to complete meiosis quiescence and male sterility [1, 49]. In addition, the inhibition of miR-871 and 880 has been shown to result in meiosis stagnation [50]. Interestingly, miR-34b-5p and miR-34-c have been shown to be important during meiosis. The miR-34b-5p effect has been illustrated to act *via* targeting Cdk6, whereas miR-34-c acts *via* the notch signaling pathway. Both miRNAs have an additional role in pachytene spermatocytes and round spermatids associated with apoptosis events [1, 18, 51]. miR-449 cluster is another regulatory miRNA found to be involved in meiotic and postmeiotic events during the later stages of spermatogenesis. It has also been shown to be crucial in apoptotic events *via* BCL2 and AFT1 [18, 46].

piRNA is the prevalent ncRNA during spermatogenesis. It has been found to account for about 17% of the ncRNA in human testis tissue and is expressed highly in immature and mature sperms [33]. Recent data in mice has indicated a selective expression of piRNA during the pre-pachytene and pachytene stages. The pre-pachytene stage is rich in transposon sequences and mainly associated with prenatal and fetal male germ cell *de novo* DNA methylation. Pre-pachytene piRNA is co-expressed with MIWI2 and/or MILI. The complex of pachytene piRNA, MIWI, and the deadenylase CAF1 join to form the pi-RISC, by which it initiates the degradation of a huge number of mRNAs in elongating spermatids. The same complex plays a protecting function depending on the state of the complex and piRNA loading [14]. TDRD is believed to act as a scaffold to recruit PIWI proteins to form germline granules, a process essential for PIWI/piRNA function during spermatogenesis [14, 52].

Research on lncRNAs has just taken up a notch with reports on their expression profile and role in experimental models. Many lncRNAs were found to express at different stages of spermatogenesis and were reported most abundant in testicular tissues when compared to other tissue types [53]. However, only a few have been characterized to carry out important functions during germ cell development. Spga-lncRNAs1 & 2 have been described as spermatogonial specific and found essential for maintaining the stemness of spermatogonia. LncRNA033862 is another important regulator of SSC self-renewal *via* activating Gfra1gene. These lncRNAs are transcribed from the antisense strand of the GDNF alpha family receptors (Gfra1) gene. It was found to be highly expressed in spermatogonia and SSCs and can be dysregulated by GDNF (Gfra1 ligand) in mouse SSCs [18, 53]. The Dmr1-related gene is an intriguing lncRNA, unintentionally discovered by Zhong [54]. The group found an unexpected size sequence at the 3′ end while trying to clone the Dmrt1 gene. Further, investigation indicated that this part sequence was coded by the Dmr gene and located on chromosome 5, therefore forming a trans-splicing RNA isoform with Dmrt1 (located on chromosome 19). This disruption to the sequence resulted in the downregulation of the Dmrt protein and was later proposed to participate in regulating germ cell development *via* switching between mitosis and meiosis [37].

In the later stages of spermatogenesis, in human, Narcolepsy candidate-region 1 gene (NLCI-C) is a lncRNA found to express in the cytoplasm of spermatogonia and early spermatocytes. The inhibition of this lncRNA has been shown to promote apoptosis, whereas its overexpression stimulated cell growth. In agreement, the study by Nishant et al., [55] demonstrated that NLC1-C promoted the proliferation of spermatogonia and spermatocytes in maturation arrest patients [53, 55]. In rodents, HongrES2 lncRNA was reported to follow a pattern of expression in the cauda region of the epididymis that matches its proposed role as a mediator of sperm maturation. This lncRNA is processed in the nucleus to a 23 bp RNA known as mil-HongrES2. The overexpression of the processed transcripts resulted in a lower sperm capacitation [53, 56]. Moreover, lncRNA-HSVIII and lncRNA- T cam1 have been suggested to participate in the regulation of transcriptional expression of the spermatocyte-specific gene [57]. Testis-specific X-linked lncRNA is another crucial lncRNA located within the X-inactivation center. It is known to be highly expressed in the pachytene spermatocytes and has been proposed to regulate meiotic division [18, 53]. lncRNA-Tug1 is among the most conserved lncRNA between humans and mice. It resides on chromosome 11 in murine and was found to be widely expressed during different developmental stages in both humans and mice. Tug1$^{-/-}$ males were presented with severe oligozoospermia with head and midpiece sperm morphological defects [53, 58]. Lnc32058, lnc98487, and lnc09522 have been identified to be involved in sperm motility-related regulatory processes [53]. Male important reproductive and fertility-related ncRNA are presented in **Table 2**.

Taken together, ncRNAs clearly, therefore, play a pivotal role during male reproduction. They are involved in SSCs differentiation, proliferation, and self-renewal. Moreover, they regulate the cell cycle, apoptosis, cell communications, and signaling pathways within the male reproductive system, and thus, ncRNAs can potentially be used as a critical biomarker for disease detection in the male reproductive tract. Although a number of database and analysis tools to support many of the potential discoveries of ncRNAs are now publicly available, functional annotations of different ncRNAs in spermatogenesis are in outset and need strong experimental evidence. The availability of the current technologies may allow us in the future to use ncRNA as a diagnostic or treatment target for male infertility.

Expression in reproductive tissues	miRNA	piRNA	lncRNAs	Function of target	Ref.
Spermatogonium	miR-10b miR-202-3p miR-34c, miR-21, miR-146a miR-182, miR-183, and clusters of miR-17-92 and miRNA 290–295. miRNAs, such as miR-221/222, miR-146	TDRD	Spga-lncRNAs1 & 2, 033862, Dmrt1, NLCI-C	SSC self-renewal Regulating germ cell development Act as scaffold to recruit PIWI proteins. Promote SSCs differentiation and self-renewal. SSCs self-renewal. Initiating spermatogonial meiosis Act to maintain SSCs in an undifferentiated state.	[11, 18, 53] [14, 18, 48] [4, 52] [18, 43] [45, 59] [14, 18, 48]
Spermatocytes	miR-469 miR-871 and miR-880, miR-34b-5p and miR-34-c		NLCI-C Xist	Transition from pachytene spermatocyte to spermatid. Regulate meiotic division. Promoted the proliferation of spermatogonia and spermatocytes. Regulate meiotic division.	[50] [50] [55] [18] [53]
Spermatid	miR-122-a, miR-23b, miR-30c, miR—690. miR-449, miR-34b-5p and miR-34-c			Cell adhesion and spermeation. Regulate round spermatids associated with apoptosis events.	[11, 18, 51] [11, 18, 51]
Sperm/Seminal plasma	miR-34c-5p, miR-122, miR-146b-5p, miR-181a, miR-509-5p, miR-513a-5p, miR-193b	piR-31,068, piR-31,098, piR-31,925, piR-43,773	Lnc32058, lnc98487, and lnc09522	Regulate sperm motility related processes.	[53]

Table 2.
Important ncRNA related to male reproduction and fertility.

2.2 Roles of noncoding RNAs in follicular development and oocyte maturation

The ovarian follicle is the principal unit of the ovary, encompassing the developing oocyte in a process called oogenesis. This is a highly precise, orchestrated process by which a small resting primordial follicle, containing the oocyte and the surrounding GCs, progressively develops into a maturing ovulatory follicle. It is a cyclic process controlled by both the follicle-stimulating hormone (FSH) and luteinizing hormone (LH). Every month, under the influence of FSH and other factors, only a few follicles are selected to grow to the preovulatory stage, after which, successive LH surges modulate the growth of only one follicle to become the dominant follicle (DF), which eventually ovulate. The rest of the growing subordinate follicles (SFs) will undergo atresia. The oocyte development and follicle maturation represent the most complex dramatic cellular proliferation, differentiation, and apoptotic pathways in females. It can only be achieved by bidirectional communication between the oocytes and GCs and is tightly regulated by a myriad network of interacting genes. Therefore, dysregulation of any of the key regulatory genes would negatively impact the fate of the developing follicle.

ncRNA posttranscriptional regulation has recently gained attention during oogenesis and ovarian functions (**Figure 2**). The core of publications focused on miRNA. The expression of miRNA has been reported to vary according to the cell type, stage, and functions [18, 59]. In mice knockout models, oocyte-specific Dicer removal caused spindle disruption and chromosome aggregation, and therefore meiosis stagnation [60]. The same experiment in GC resulted in accelerated use of follicle reserve, increased number of atretic follicles, and disruption of many follicular developmental key regulatory genes [18, 61, 62]. Several studies have indicated the key involvement of miRNAs in GC survival, proliferation, and differentiation, as well as steroidogenesis. miR-224, under the influence of transforming growth factor-beta1, has been shown to play a role in regulating GCs proliferation and function [1, 18]. Moreover, FSH was reported to modulate the expression of miR-29a and 30d in cultured GCs. Furthermore, the LH surge appeared to promote miR-21, miR-212, and miR132 expressions in GCs. The overexpression of miR-143 inhibited primordial follicle formation in 15.5 dpc murine ovaries [18].

piRNAs have been found to be abundantly expressed in the ovaries. In humans, MILI was specifically localized to fetal oocytes, whereas MIWI and MILI were found in adult ovaries. piRNA has been described to be highly expressed in the oocytes of rhesus monkeys, bovine, as well as in humans [63]. However, their function is undescribed yet. PIWI proteins were localized to mouse oocytes, yet their loss did not

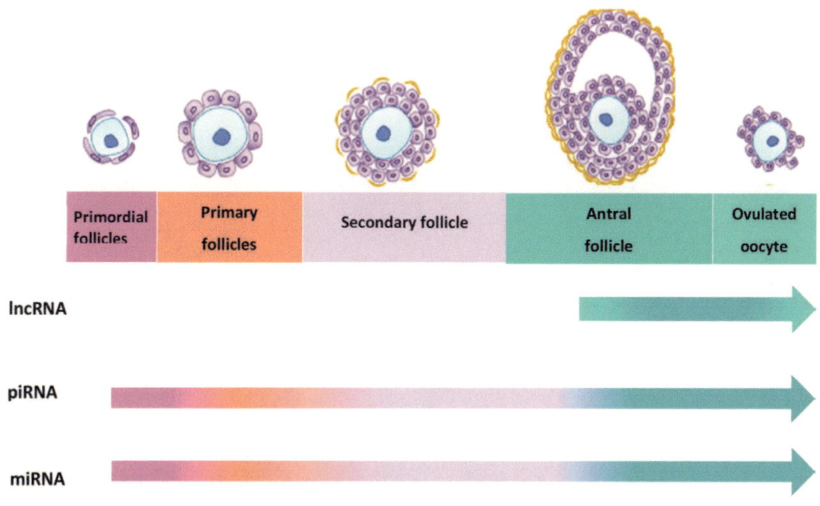

Figure 2.
Characterizations expression of miRNAs, piRNA, and lncRNAs during the different stages of oogenesis [12, 18].

affect their fertility. The PIWIL1 knockout model was interestingly able to generate mature fertilizable oocytes, nevertheless, arrested at the two-cell stage. The author indicated, as a result, the accumulation of transposons in the oocytes and zygotic gene activation failure [64].

lncRNA-H19 was the first discovered functional lncRNA in the female reproductive tract. They were considered as a large functional RNA sequence with no coding proteins. H19 was described to be significantly expressed in endoderm and mesoderm-derived tissues. It has been shown to involve in the modulation of a co-expressing net of imprinting genes [65]. lncRNAs were recently found to play a role in regulating oocyte meiosis. It has been demonstrated that lncRNA Y00062 may be involved in NEK7 modulation, a known crucial gene for spindle assembly and mitosis. Moreover, lncRNA ENST00000502390 was reported to overexpress in

Expression in reproductive tissues	miRNA	piRNA	lncRNAs	Function of target	Ref.
Folliculogenesis	miR-143		H19, Xist, AK124742, Amhr2		[67]
Granulosa cells	miR-133b, miR-98, miR-29a, miR-23a, miR-15a, miR-10a, miR-145, miR-151, miR-22-3p, miR-146a, miR-21, miR-106a, miR-24, miR-320a, miR-1275, miR-125a/b, miR-100, miR-224, miR-30d, miR—132, miR-212		H19, Let-7, Let -7 g, Neat1,ENST0000050239	Steroidogenesis, GC proliferation and apoptosis	[11]
Oocyte	miR-10a, miR-184, miR-100. miR-20a, miR-15a, miR-602		lncRNA Y00062, ENST0000050239	spindle assembly and mitosis Oocyte reprogramming, Repression nuclear receptors; regulation of oocyte specific genes expression. Regulation of cell growth and division.	[66] [68]

Table 3.
Important ncRNA related to female reproduction and fertility.

poor-quality cumulus cells (CCs) and is associated with elongation of very long-chain fatty acid protein 5 (ELOVL5). ELOVL5 is known to involve in modulating oocyte maturation and ovulation and in the biosynthesis of highly unsaturated fatty acids, thus ENST00000502390 may participate in the same processes [66]. Thus, strengthening the evidence for the role of these ncRNAs in the oocyte maturation processes. Some examples of known female reproductive and fertility-related ncRNA are presented in **Table 3**.

Overall, these data demonstrate the expression of the different ncRNAs within the various organs of the female reproductive tract, where they play an important role in modulating crucial cellular pathways for proper functioning. However, it is still a vague area of research and represents a great challenge in reproductive medicine. Upcoming studies, nevertheless, can provide a new insight into utilizing these ncRNAs as biomarkers and therapeutic targets for fertility disorders.

3. The role of noncoding RNAs in some reproductive disorders

The development of gametes is extremely complex, requiring both the sperm and egg to passage through several stages to eventually form the mature sperm in males, and the maturing follicle finally discharges the mature cumulus enveloped oocyte in females. This process is tightly controlled and the interruption of the sequence at any stage can impact one's fertility. ncRNAs play a crucial part of all these processes.

3.1 The role of ncRNAs in male reproductive disorders

Male infertility is evident in about 40–50% of all infertility cases; however, the research into this topic is relatively spars. Molecular studies in the field of male fertility are important and can unveil the etiology of infertility. ncRNA dysregulation has been shown to implicate in male infertility such as prostatic cancer (PCa), varicocele, and sperm abnormalities.

3.1.1 Prostatic cancer (PCA)

PCa is one of the complex types of malignancies and is ranked as the second most lethal cancer type in men. A number of ncRNA transcripts have been identified and associated with PCa. These transcripts have been named as prostate cancer-associated ncRNA transcripts (PCATs). lncRNA PCAT-1, for example, has been found to be highly expressed in the cytoplasm and in low amount in the nucleus of PCa cells [10, 69]. In addition, the co-expression of lncRNA PCAT-1 with miR-34a and miR-3667-3 has been described as responsible for increased prostate tumor cell proliferation. This was achieved *via* upregulating the c-Myc protein, a crucial protein for cell cycle progression [10, 70]. Testis-specific transcript Y-linked 15 (TTTY15) lncRNA has been reported to be highly expressed in PCa patient samples when compared to the control group. The authors showed that TTTY15 acts as a sponge to let-7 miRNA to upregulate CDK6 and FN1 in those patients [10]. miR-648 has been identified as a novel diagnostic biomarker for PCa. The upregulation of miR-181a, miR-122, miR-181c, miR-193a, miR-193b, and miR-125a-b have been reported in PCa tissue samples, whereas miR-145, miR-125, and lect-7c were found dysregulated [71–73]. novel_ pir382289, novel_ pir158533, novel_ pir349843, and has_pir_002468 have been associated with PCa development [74].

3.1.2 Varicocele (VC)

It is identified by an abnormal swelling of the pampiniform venous plexus within the scrotum. This condition can cause venous pressure and increased temperature, hypoxia-related oxidative stress, and apoptosis, leading to testicular dysfunction and impact sperm function and quality [10, 75]. Several studies indicated miRNA as an important player in the development of VC. A study by Ou and colleagues demonstrated that caspase-3 sperm cell apoptosis has been induced by miR-210-3P in patients with VC [76]. The same group further assessed the relationship between VC development and the expression profile of rno-miRNA-210, rno-miR-190a, rno-miR-6316, and rno-miR-135bb-5p. These findings provided evidence indicating the upregulation of these miRNA in VC patients' samples [76]. Another study compared the expression of seminal miRNAs and oxidative stress-related apoptotic markers in VC patients. The group reported the downregulation of miR-122, miR-34c5, and miR-181a in infertile patients with VC [77].

The overexpression of lncRNA gadd7 (growth arrested DNA-damage inducible gene 7) has been associated with DNA damage and inhibition of cell growth. Increased expression of lncRNA gadd 7 has been studied in VC semen samples and was found to negatively impact sperm count [78].

3.1.3 Sperm abnormalities

Sperm abnormalities are a fundamental cause of male infertility such as obstructive and nonobstructive azoospermia, oligozoospermia, asthenozoospermia, and teratozoospermia [10]. miRNAs dysregulation has been linked to Sertoli cell only (SCO) syndrome, asthenozoospermia, germ cell arrest, and mixed atrophy (MA) [14]. Single-nucleotide polymorphisms have been localized in miRNA-binding sites of male fertility-associated genes, indicating a role of miRNAs in male infertility. In addition, Dorsha and Dicer SNPs have been shown to affect sperm quality, which further supports their role in male infertility.

Recent evidence demonstrated that multiple SNPs from human piwi genes are associated with increased risk of oligozoospermia or spermatogenic failure [79]. Moreover, spermatogenesis disruption can be due to PIWIL1/PIWIL2 allele-specific DNA methylation [80].

Differential expression of lncRNA in oligoasthenozoospermic and asthnozoospermic patients has been associated with spermatogenesis and sperm function. HOTAIR-lncRNA down-regulation has been shown to be related to decreased nuclear factor erythroid 22-related factor 2 (NRF2) gene expression. The expression levels of the NRF2 gene have been related to antioxidant gene expression and sperm quality. Thus, suggesting a protective function of HOTAIR against antioxidant sperm-damaging activity [14]. The level of expression of lncRNA32058, lnc98487, and lncRNA09522 have been reported to significantly impact sperm motility [81].

To date, thousands of ncRNAs have been identified to be associated with different male reproductive disorders. However, the current mechanisms and molecular pathways modulating these processes are still underway. Unveiling the link between the differential expression and functions of these ncRNAs and each male reproductive disease can identify needed biomarkers and support the development of new approaches to treat these conditions.

3.2 The role of ncRNAs in female reproductive disorders

ncRNAs have recently been explored as molecular markers for female reproductive-related disorders such as polycystic ovarian syndrome (PCOS), premature ovarian failure (POF), and repeated implantation failure (RIF).

3.2.1 Polycystic ovarian syndrome (PCOS)

PCOS is a common systematic endocrine disorder affecting women at their reproductive age. It is characterized by the dysregulation of androgens, which makes it a prevalent cause of anovulation. miRNA-mediated regulation of androgens has been explored recently in women with PCOS. The study results suggested an inverse relationship between the expression levels of serum miR-23a and serum testosterone [82]. Moreover, the overexpression of three miRNAs: miR-222, miR-30c, and miR-146a has been indicated in patients with PCOS when compared to controls [18]. Circulating miRNAs can be profiled in plasma, serum, and follicular fluid. A recent study demonstrated the upregulation of miR-32, miR-18b, miR-135a, miR34c, and miR-9 in the follicular fluid of PCOS patients. On the other hand, miR-320, and miR-132 were downregulated in the same group [68].

The role of lncRNA in the follicular fluid and somatic cells and the development of PCOS have recently been explored. For example, two GC and CC specific lncRNA, PWRN2 and HCG26 have been associated with PCOS. The overexpression of PWRN2 and its target gene ATP6V1G3 have been proposed to act *via* decreasing the follicular PH, and therefore oocyte dysplasia in these patients. HCG26 lncRNA has been shown to play a role in the regulation of cell proliferation and steroidogenesis. Therefore, alteration expression of this lncRNA in GC may contribute to the pathogenicity of this disorder.

3.2.2 Premature ovarian failure (POF)

POF is a multifactorial disorder characterized by hypoestrogenism and hypogonadism in women under the age of 40. Like PCOS, ncRNAs have been explored as epigenetic markers for the disease. The overexpression of circulating miRNAs, such as miR-126, miR-125b-2, miR-23a, miR-27a, miR146a, miR-139-3p and downregulation of miR-144, let-7c, and miR-22-3p, have been noted in patients with POF [68, 83].

Data on lncRNAs in POF is still quite limited and requires further research. The correlation between the expression profile of lncRNAs FMR4 and FMR6 and POF has quite recently been studied. The authors observed a significantly negative association between the expression level of FMR6 in GCs and the number of oocytes retrieved. Moreover, in mice, cyclophosphamide *via* upregulated lncRNA-Meg3-p53-p66Shc pathway has been proposed to induce ovarian injury and POF [67, 84].

3.2.3 Ovarian cancer

Errors during the developmental process and the dysregulation of molecular pathways can give rise to different types of cancer. Several miRNAs have been associated with the manifestation of ovarian cancers and others appear to act as cancer suppressors [83]. LIN28 protein is specifically expressed in ESCs and known as a principal modulator of human embryonic stem cells (hESC) pluripotency. It has been reported to function as an oncogene to promote tumor progression. Zhong and colleagues demonstrated that the action of this protein may be suppressed in hESCs and cancer cells by the expression of four miRNAs, miR-9, miR-125, miR-30, and let-7 [54].

HOTAIR lncRNA has been shown to be highly expressed in endothelial ovarian tissues and serous ovarian cancers [85]. Another lncRNA, Xist, is a transcript of an inactive X chromosome and has been found to express in very small amounts in ovarian cancer cell lines [86]. Furthermore, H19 lncRNA is one of the first lncRNAs to express in epithelial ovarian cancers. MEG3 lncRNA has been reported to suppress tumor cell development and progression in various types of cancers, including ovarian cancer [85, 87].

It is indeed well documented that ncRNAs play an important role in female reproductive disease-related conditions. However, there is still a challenge in determining their specific regulatory contribution in relation to the disease state. Further research and studies are required to understand the exact mechanisms of their involvement and how they can be targeted for therapeutic interventions. Further, studies are needed to further provide evidence and expand our database for the diagnostic and therapeutic targets for female-related-reproductive disorders.

4. Potential applicability of noncoding RNAs as a diagnostic noninvasive biomarker and as a predictor of oocyte/embryo quality and reproductive disorders

Growing evidence indicates a crucial role of ncRNAs in infertility and reproductive health-related disorders. The detection of these ncRNAs can be easily isolated from various human biological fluids such as serum, plasma, seminal plasma, and follicular fluids, which makes them a reliable tool for infertility-related care.

At present, the quality of retrieved oocytes and developing embryos during the assisted reproductive program can only be assessed morphologically or *via* an invasive aggressive embryo biopsy. Thus, improving techniques to isolate and evaluate ncRNAs in spent culture medium can provide an interesting noninvasive tool to assess those embryos. In addition, *in vivo* or *in vitro* aging of postovulatory oocyte is a challenging scenario. Often, those oocytes are presented with chromosomal abnormalities, polyspermy, lower fertilization, and abnormal embryo development [17, 88]. It has been reported that these abnormalities are due to various cellular and molecular changes occurring within the aging oocytes. These changes include intracytoplasmic adenosine triphosphate (ATP) and antioxidant GSH reduction, increased reactive oxygen species (ROS), morphological and organelle alterations, reduction in expression of antiapoptotic factor Bcl-2, Ca^{2+} dysregulation, and increased apoptosis [17, 89]. Consequently, oxidative stress on these oocytes leads to the trigger of a cascade of events that results in postovulatory oocytes aging. ncRNAs have been shown to be involved in controlling these processes and therefore, understanding the regulatory mechanism can aid the prevention and maybe the development of a supplementation to protect these oocytes *in vitro*. Interestingly, the use of hCG and vitamin C supplementation in the culturing medium has been shown to alter the expression profile of miRNAs in murine GCs and oocytes during *in vitro* growth [83, 90]. In addition, current data provides some understanding regarding the involvement of ncRNAs in hESCs, which literally can comprehend some human reproductive entity development, and therefore some cancer manifestation mysteries. Thus, enhance and support reproductive-related cancer treatment protocols.

Early diagnosis can aid the management and treatment of various reproductive disorders. Yet, the profiling of ncRNAs from the culturing medium and biological fluids is still challenging. The small amount of ncRNAs in these samples creates some technical difficulties, including material expression variations, handling procedures may introduce contamination or cause RNA degradation, and the accuracy of the technology used for isolation, all of which can result in diversity in the reported profiles. For example, high-throughput sequencing analysis provides large data of short sequence reads, which often contain contaminant sequences in addition to the other primer sequencing. Therefore, the development of highly advanced and complex algorithms is needed to clear out any inaccurate reads and produce more robust data. In addition, the intricate ncRNA networks that are responsible for regulating reproductive health can make it extremely difficult to understand how they function. The amount of complexity involved in these molecular pathways can be quite challenging to decipher. In fact, it raises questions about wither it is the actual cause or part of the targeted pathway associated with the pathological condition. It has, however, been clearly shown that ncRNAs may provide a great potential for its use as a biomarker for the diagnosis and treatment of reproductive disorders.

5. Conclusion

Growing evidence indicates a crucial role of ncRNAs in epigenetic modulation and regulation of infertility-related disorders. Recent advancements in molecular testing and informatics approaches are well established and have eased ncRNA research. Discoveries of more novel species on ncRNAs are ongoing and can contribute to unveiling the "dark matter" of the genome, to further knowledge of ours in infertility-related disorders.

Author details

Rana Alhamdan
King Faisal Specialist Hospital and Research Center, Riyadh, Saudi Arabia

*Address all correspondence to: amsf199705@hotmail.co.uk

References

[1] He C, Wang K, Gao Y, Wang C, Li L, Liao Y, et al. Roles of noncoding RNA in reproduction. Frontiers in Genetics. 2021;**12**:777510

[2] Zorina ZA, Obozova TA. New data on the brain and cognitive abilities of birds. Zoologichesky Zhurnal. 2011;**90**:784-802

[3] Wang KC, Chang HY. Molecular mechanisms of long noncoding RNAs. Molecular Cell. 2011;**43**:904-914

[4] Imbar T, Galliano D, Pellicer A, Laufer N. Introduction: Micrornas in human reproduction: Small molecules with crucial regulatory roles. Fertility and Sterility. 2014;**101**:1514-1515

[5] Bouckenheimer J, Assou S, Riquier S, Hou C, Philippe N, Sansac C, et al. Long non-coding RNAs in human early embryonic development and their potential in art. Human Reproduction Update. 2017;**23**:19-40

[6] Dunham I, Kundaje A, Aldred SF, Collins PJ, Davis CA, Doyle F, et al. An integrated encyclopedia of DNA elements in the human genome. Nature. 2012;**489**:57-74

[7] Lee H, Zhang Z, Krause HM. Long noncoding RNAs and repetitive elements: Junk or intimate evolutionary partners? Trends in Genetics. 2019;**35**:892-902

[8] Rinn JL, Chang HY. Genome regulation by long noncoding RNAs. Annual Review of Biochemistry. 2012;**81**:145-166

[9] Wang Z, Gerstein M, Snyder M. RNA-Seq: A revolutionary tool for transcriptomics. Nature Reviews Genetics. 2009;**10**:57-63

[10] Tezerjani MD, Kalantar SM. Unraveling the dark matter, long non-coding RNAs, in male reproductive diseases: A narrative review. International Journal of Reproductive Biomedicine. 2020;**18**:921

[11] Zhang R, Wesevich V, Chen Z, Zhang D, Kallen AN. Emerging roles for noncoding RNAs in female sex steroids and reproductive disease. Molecular and Cellular Endocrinology. 2020;**518**:110875

[12] Zhu Q, Kirby JA, Chu C, Gou L-T. Small noncoding RNAs in reproduction and infertility. Biomedicine. 2021;**9**:1884

[13] Iwasaki YW, Siomi MC, Siomi H. PIWI-interacting RNA: Its biogenesis and functions. Annual Review of Biochemistry. 2015;**84**:405-433

[14] Gou LT, Dai P, Liu MF. Small noncoding RNAs and male infertility. Wiley Interdiscip Rev RNA. 2014;**5**:733-745

[15] Sabry R, Yamate J, Favetta L, Lamarre J. MicroRNAs: Potential targets and agents of endocrine disruption in female reproduction. Journal of Toxicologic Pathology. 2019;**32**:213-221

[16] Zhu H-L, Chen Y-Q, Zhang Z-F. Downregulation of lncRNA Zfas1 and upregulation of microRNA-129 repress endocrine disturbance, increase proliferation and inhibit apoptosis of ovarian granulosa cells in polycystic ovarian syndrome by downregulating HMGB1. Genomics. 2020;**112**:3597-3608

[17] Gebremedhn S, Ali A, Hossain M, Hoelker M, Salilew-Wondim D, Anthony RV, et al. MicroRNA-mediated gene regulatory mechanisms in mammalian female reproductive health.

International Journal of Molecular Sciences. 2021;**22**:938

[18] Robles V, Valcarce DG, Riesco MF. Non-coding RNA regulation in reproduction: Their potential use as biomarkers. Non-coding RNA Research. 2019;**4**:54-62

[19] Le Thomas A, Tóth KF, Aravin AA. To be or not to be a piRNA: Genomic origin and processing of piRNAs. Genome Biology. 2014;**15**:204

[20] Huang Y, Bai JY, Ren HT. piRNA biogenesis and its functions. Russian Journal of Bioorganic Chemistry. 2014;**40**:293-299

[21] Wu X, Pan Y, Fang Y, Zhang J, Xie M, Yang F, et al. The biogenesis and functions of piRNAs in human diseases. Molecular Therapy - Nucleic Acids. 2020;**21**:108-120

[22] Chen D, Zhang Z, Chen B, Ji D, Hao Y, Zhou P, et al. Altered microRNA and Piwi-interacting RNA profiles in cumulus cells from patients with diminished ovarian reserve†. Biology of Reproduction. 2017a;**97**:91-103

[23] Ketting RF. The many faces of RNAi. Developmental Cell. 2011;**20**:148-161

[24] Kabayama Y, Toh H, Katanaya A, Sakurai T, Chuma S, Kuramochi-Miyagawa S, et al. Roles of MIWI, MILI and PLD6 in small RNA regulation in mouse growing oocytes. Nucleic Acids Research. 2017;**45**:5387-5398

[25] Saxe JP, Chen M, Zhao H, Lin H. TDRKH is essential for spermatogenesis and participates in primary piRNA biogenesis in the germline. The EMBO Journal. 2013;**32**:1869-1885

[26] St Laurent G, Wahlestedt C, Kapranov P. The landscape of long noncoding RNA classification. Trends in Genetics. 2015;**31**:239-251

[27] Derrien T, Johnson R, Bussotti G, Tanzer A, Djebali S, Tilgner H, et al. The gencode V7 catalog of human long noncoding RNAs: Analysis of their gene structure, evolution, and expression. Genome Research. 2012;**22**:1775-1789

[28] Mattick JS, Amaral PP, Carninci P, Carpenter S, Chang HY, Chen L-L, et al. Long non-coding RNAs: Definitions, functions, challenges and recommendations. Nature Reviews Molecular Cell Biology. 2023;**24**:430-447

[29] Ma L, Bajic VB, Zhang Z. On the classification of long non-coding RNAs. RNA Biology. 2013;**10**:925-933

[30] Hsiao K-Y, Sun HS, Tsai S-J. Circular RNA – New member of noncoding RNA with novel functions. Experimental Biology and Medicine. 2017;**242**:1136-1141

[31] Feng Y, Hu X, Zhang Y, Zhang D, Li C, Zhang L. Methods for the study of long noncoding RNA in cancer cell signaling. Methods in Molecular Biology. 2014;**1165**:115-143

[32] Williams Z, Morozov P, Mihailovic A, Lin C, Puvvula PK, Juranek S, et al. Discovery and characterization of piRNAs in the human fetal ovary. Cell Reports. 2015;**13**:854-863

[33] Aliakbari F, Eshghifar N, Mirfakhraie R, Pourghorban P, Azizi F. Coding and non-coding RNAs, as male fertility and infertility biomarkers. International Journal of Fertility & Sterility. 2021;**15**:158-166

[34] Balas MM, Johnson AM. Exploring the mechanisms behind long noncoding RNAs and cancer. Non-coding RNA Research. 2018;**3**:108-117

[35] Luo LF, Hou CC, Yang WX. Small non-coding RNAs and their associated proteins in spermatogenesis. Gene. 2016;**578**:141-157

[36] Mukherjee A, Koli S, Reddy KV. Regulatory non-coding transcripts in spermatogenesis: Shedding light on 'dark matter'. Andrology. 2014;**2**:360-369

[37] Luk AC-S, Chan W-Y, Rennert OM, Lee T-L. Long noncoding RNAs in spermatogenesis: Insights from recent high-throughput transcriptome studies. Reproduction. 2014;**147**:R131-R141

[38] Moritoki Y, Hayashi Y, Mizuno K, Kamisawa H, Nishio H, Kurokawa S, et al. Expression profiling of microrna in cryptorchid testes: miR-135a contributes to the maintenance of spermatogonial stem cells by regulating Foxo1. The Journal of Urology. 2014;**191**:1174-1180

[39] Yan N, Lu Y, Sun H, Tao D, Zhang S, Liu W, et al. A microarray for microRNA profiling in mouse testis tissues. Reproduction. 2007;**134**:73-79

[40] Wu Q, Song R, Ortogero N, Zheng H, Evanoff R, Small CL, et al. The RNase III enzyme DROSHA is essential for microRNA production and spermatogenesis. The Journal of Biological Chemistry. 2012;**287**:25173-25190

[41] Papaioannou MD, Nef S. Micrornas in the testis: Building up male fertility. Journal of Andrology. 2010;**31**:26-33

[42] Papaioannou MD, Pitetti JL, Ro S, Park C, Aubry F, Schaad O, et al. Sertoli cell dicer is essential for spermatogenesis in mice. Developmental Biology. 2009;**326**:250-259

[43] Yu M, Mu H, Niu Z, Chu Z, Zhu H, Hua J. miR-34c enhances mouse spermatogonial stem cells differentiation by targeting Nanos2. Journal of Cellular Biochemistry. 2014;**115**:232-242

[44] He Z, Jiang J, Kokkinaki M, Tang L, Zeng W, Gallicano I, et al. miRNA-20 and miRNA-106a regulate spermatogonial stem cell renewal at the post-transcriptional level via targeting STAT3 and Ccnd1. Stem Cells. 2013;**31**:2205-2217

[45] Chen J, Cai T, Zheng C, Lin X, Wang G, Liao S, et al. MicroRNA-202 maintains spermatogonial stem cells by inhibiting cell cycle regulators and RNA binding proteins. Nucleic Acids Research. 2016;**45**:4142-4157

[46] Chen X, Li X, Guo J, Zhang P, Zeng W. The roles of microRNAs in regulation of mammalian spermatogenesis. Journal of Animal Science and Biotechnology. 2017b;**8**:35

[47] Niu B, Wu J, Mu H, Li B, Wu C, He X, et al. miR-204 regulates the proliferation of dairy goat spermatogonial stem cells via targeting to Sirt1. Rejuvenation Research. 2016;**19**:120-130

[48] Huszar JM, Payne CJ. MicroRNA 146 (miR146) modulates spermatogonial differentiation by retinoic acid in Mice1. Biology of Reproduction. 2013;**88**:15/1-10

[49] Gao H, Wen H, Cao C, Dong D, Yang C, Xie S, et al. Overexpression of microRNA-10a in germ cells causes male infertility by targeting Rad51 In mouse and human. Frontiers in Physiology. 2019;**10**

[50] Ota H, Ito-Matsuoka Y, Matsui Y. Identification of the X-linked germ cell specific miRNAs (Xmirs) and their functions. PLoS One. 2019;**14**:E0211739

[51] Wang X, Zhang P, Li L, Che D, Li T, Li H, et al. miRNA editing landscape reveals miR-34c regulated spermatogenesis through structure

and target change in pig and mouse. Biochemical and Biophysical Research Communications. 2018;**502**:486-492

[52] Meikar O, Da Ros M, Korhonen H, Kotaja N. Chromatoid body and small RNAs in male germ cells. Reproduction. 2011;**142**:195-209

[53] Joshi M, Rajender S. Long non-coding RNAs (lncRNAs) in spermatogenesis and male infertility. Reproductive Biology and Endocrinology. 2020;**18**:103

[54] Zhong X, Li N, Liang S, Huang Q, Coukos G, Zhang L. Identification of microRNAs regulating reprogramming factor LIN28 in embryonic stem cells and cancer cells. The Journal of Biological Chemistry. 2010;**285**:41961-41971

[55] Nishant KT, Ravishankar H, Rao MR. Characterization of a mouse recombination hot spot locus encoding a novel non-protein-coding RNA. Molecular and Cellular Biology. 2004;**24**:5620-5634

[56] Ni MJ, Hu ZH, Liu Q, Liu MF, Lu MH, Zhang JS, et al. Identification and characterization of a novel non-coding RNA involved in sperm maturation. PLoS One. 2011;**6**:E26053

[57] Anguera MC, Ma W, Clift D, Namekawa S, Kelleher RJ III, Lee JT. Tsx produces a long noncoding RNA and has general functions in the germline, stem cells, and brain. PLoS Genetics. 2011;7:E1002248

[58] Lewandowski JP, Dumbović G, Watson AR, Hwang T, Jacobs-Palmer E, Chang N, et al. The Tug1 lncRNA locus is essential for male fertility. Genome Biology. 2020;**21**:237

[59] Hossain MM, Ghanem N, Hoelker M, Rings F, Phatsara C, Tholen E, et al. Identification and characterization of miRNAs expressed in the bovine ovary. BMC Genomics. 2009;**10**:443

[60] Murchison EP, Stein P, Xuan Z, Pan H, Zhang MQ, Schultz RM, et al. Critical roles for dicer in the female germline. Genes & Development. 2007;**21**:682-693

[61] Gonzalez G, Behringer RR. Dicer is required for female reproductive tract development and fertility in the mouse. Molecular Reproduction and Development. 2009;**76**:678-688

[62] Hong X, Luense LJ, Mcginnis LK, Nothnick WB, Christenson LK. Dicer1 is essential for female fertility and normal development of the female reproductive system. Endocrinology. 2008;**149**:6207-6212

[63] Roovers EF, Rosenkranz D, Mahdipour M, Han CT, He N, De Sousa C, et al. Piwi proteins and piRNAs in mammalian oocytes and early embryos. Cell Reports. 2015;**10**:2069-2082

[64] Hasuwa H, Iwasaki YW, Au Yeung WK, Ishino K, Masuda H, Sasaki H, et al. Production of functional oocytes requires maternally expressed Piwi genes and piRNAs in golden hamsters. Nature Cell Biology. 2021;**23**:1002-1012

[65] Gabory A, Jammes H, Dandolo L. The H19 locus: Role of an imprinted non-coding RNA in growth and development. BioEssays. 2010;**32**:473-480

[66] Warzych E, Pawlak P, Pszczola M, Cieslak A, Madeja ZE, Lechniak D. Interactions of bovine oocytes with follicular elements with respect to lipid metabolism. Animal Science Journal. 2017;**88**:1491-1497

[67] Wang XY, Qin YY. Long non-coding RNAs in biology and female reproductive

disorders. Frontiers in Bioscience (Landmark Edition). 2019;**24**:750-764

[68] Kamalidehghan B, Habibi M, Afjeh SS, Shoai M, Alidoost S, Almasi Ghale R, et al. The importance of small non-coding RNAS in human reproduction: A review article. The Application of Clinical Genetics. 2020;**13**:1-11

[69] Prensner JR, Iyer MK, Balbin OA, Dhanasekaran SM, Cao Q, Brenner JC, et al. Transcriptome sequencing across a prostate cancer cohort identifies PCAT-1, an unannotated lincRNA implicated in disease progression. Nature Biotechnology. 2011;**29**:742-749

[70] Prensner JR, Chen W, Han S, Iyer MK, Cao Q, Kothari V, et al. The long non-coding RNA PCAT-1 promotes prostate cancer cell proliferation through cMYc. Neoplasia. 2014;**16**:900-908

[71] Ozen M, Creighton C, Ozdemir M, Ittmann M. Widespread deregulation of microRNA expression in human prostate cancer. Oncogene. 2008;**27**:1788-1793

[72] Walter BA, Valera VA, Pinto PA, Merino MJ. Comprehensive microRNA profiling of prostate cancer. Journal of Cancer. 2013;**4**:350-357

[73] Zhang W, Zang J, Jing X, Sun Z, Yan W, Yang D, et al. Identification of candidate miRNA biomarkers from miRNA regulatory network with application to prostate cancer. Journal of Translational Medicine. 2014;**12**:66

[74] Peng Q, Chiu PK-F, Wong CY-P, Cheng CK-L, Teoh JY-C, Ng C-F. Identification of piRNA targets in urinary extracellular vesicles for the diagnosis of prostate cancer. Diagnostics. 2021;**11**:1828

[75] Fretz PC, Sandlow JI. Varicocele: Current concepts in pathophysiology, diagnosis, and treatment. The Urologic Clinics of North America. 2002;**29**:921-937

[76] Ou N, Song Y, Xu Y, Yang Y, Liu X. Identification and verification of hub microRNAs in varicocele rats through high-throughput sequencing and bioinformatics analysis. Reproductive Toxicology. 2020;**98**:189-199

[77] Mostafa T, Rashed LA, Nabil NI, Osman I, Mostafa R, Farag M. Seminal miRNA relationship with apoptotic markers and oxidative stress in infertile men with varicocele. BioMed Research International. 2016;**2016**:4302754

[78] Brookheart RT, Michel CI, Listenberger LL, Ory DS, Schaffer JE. The non-coding RNA Gadd7 is a regulator of lipid-induced oxidative and endoplasmic reticulum stress. Journal of Biological Chemistry. 2009;**284**:7446-7454

[79] Gu A, Ji G, Shi X, Long Y, Xia Y, Song L, et al. Genetic variants in Piwi-interacting RNA pathway genes confer susceptibility to spermatogenic failure In a Chinese population. Human Reproduction. 2010;**25**:2955-2961

[80] Friemel C, Ammerpohl O, Gutwein J, Schmutzler AG, Caliebe A, Kautza M, et al. Array-based DNA methylation profiling in male infertility reveals allele-specific DNA methylation in Piwil1 and Piwil2. Fertility and Sterility. 2014;**101**:1097, E1-1103

[81] Zhang X, Zhang P, Song D, Xiong S, Zhang H, Fu J, et al. Expression profiles and characteristics of human lncRNA in normal and asthenozoospermia sperm. Biology of Reproduction. 2019;**100**:982-993

[82] Xiong W, Lin Y, Xu L, Tamadon A, Zou S, Tian F, et al. Circulatory microRNA 23a and microRNA 23b and polycystic ovary syndrome (PCOS): The effects of body mass index and sex hormones in an eastern Han Chinese population. Journal of Ovarian Research. 2017a;**10**

[83] Virant-Klun I, Ståhlberg A, Kubista M, Skutella T. MicroRNAs: From female fertility, germ cells, and stem cells to cancer in humans. Stem Cells International. 2016;**2016**:3984937

[84] Xiong Y, Liu T, Wang S, Chi H, Chen C, Zheng J. Cyclophosphamide promotes the proliferation inhibition of mouse ovarian granulosa cells and premature ovarian failure by activating the lncRNA-Meg3-P53-P66shc pathway. Gene. 2017b;**596**:1-8

[85] Zhan L, Li J, Wei B. Long non-coding RNAs in ovarian cancer. Journal of Experimental & Clinical Cancer Research. 2018;**37**:120

[86] Kawakami T, Zhang C, Taniguchi T, Kim CJ, Okada Y, Sugihara H, et al. Characterization of loss-of-inactive X in Klinefelter syndrome and female-derived cancer cells. Oncogene. 2004;**23**:6163-6169

[87] Lyu Y, Lou J, Yang Y, Feng J, Hao Y, Huang S, et al. Dysfunction of the WT1-MEG3 signaling promotes AML leukemogenesis via P53-dependent and -independent pathways. Leukemia. 2017;**31**:2543-2551

[88] Tarín JJ, Pérez-Albalá S, Cano A. Consequences on offspring of abnormal function in ageing gametes. Human Reproduction Update. 2000;**6**:532-549

[89] Takahashi T, Igarashi H, Amita M, Hara S, Matsuo K, Kurachi H. Molecular mechanism of poor embryo development in postovulatory aged oocytes: Mini review. The Journal of Obstetrics and Gynaecology Research. 2013;**39**:1431-1439

[90] Kim YJ, Ku SY, Rosenwaks Z, Liu HC, Chi SW, Kang JS, et al. MicroRNA expression profiles are altered by gonadotropins and vitamin C status during in vitro follicular growth. Reproductive Sciences. 2010;**17**:1081-1089

Chapter 6

Functional Role of MicroRNAs in Embryogenesis

Muhammad Babar Khawar, Huma Rasheed, Ali Afzal, Muddasir Hassan Abbasi and Nadeem Sheikh

Abstract

This book chapter will provide an overview of the functional role of microRNAs (miRNAs) in embryogenesis. A brief introduction to embryogenesis and emphasis on the importance of miRNAs in gene regulation will be provided. The biogenesis and mechanism of action of miRNAs will be discussed in detail with a focus on the importance of miRNA-mRNA interaction in gene regulation. The chapter will then delve into the role of miRNAs in early embryonic development, including their importance in the establishment of the three germ layers, cell proliferation, differentiation, and apoptosis during embryogenesis. The role of miRNAs in organogenesis and tissue differentiation, specifically the formation of specific organs such as the heart, lung, liver, and brain, will also be discussed. The chapter will conclude by examining the dysregulation of miRNAs in embryonic development and disease, including teratogenicity, developmental disorders, and developmental cancer. The chapter will summarize the functional roles of miRNAs in embryogenesis and will offer future perspectives and potential therapeutic applications of miRNAs in embryonic development and disease.

Keywords: miRNAs, embryogenesis, gene regulation, organogenesis, development

1. Introduction

Small noncoding RNAs, also called miRNAs, which have a length of about 22 nucleotides, control the expression of genes at the posttranscriptional stage by selecting mRNAs for cleavage or transcriptional repression [1]. The majority of embryonic alterations can be connected to the genome's activity [2]. According to mouse studies, genomic data within preimplantation embryos undergoes wave-like alterations related with maternal transcriptome disintegration and zygote genome activation (ZGA) [3]. Despite the fact that miRNA regulates transcription, it is still unclear how it affects the removal of the maternal genome. According to studies, mouse oocytes and early preimplantation embryos exhibit usually decreased miRNA function [4]. The timing of extensive maternal gene breakdown is consistent with the inhibition of miRNA activity occurring before the two-cell stage. The expression of miRNA is revived, and the repression is lifted after the two-cell stage [5]. miRNA is short RNA molecules that regulate the translation and transcription.

The formation of miRNA is done by folding of small hairpin structure by the transcription of RNA polymerase II translation repression, and mRNA degradation is brought on by the synthesis of miRNAs in the nucleus as primary (pri)-miRNA and pre-miRNA, mature miRNA duplexes, RNA-induced silencing complex (RISC) strand-mediated complex, and complementary mRNA sequence synthesis (in the cytoplasm). Before implantation, a fertilized egg undergoes continuous cleavage to create an implantable embryo in a process known as embryogenesis. There are several factors, which characterize embryonic development, including the breakdown of maternal stores, activation of minor and major embryonic genomes, breaking down of the embryo and developing the embryonic cavity, and proliferation of embryonic cells into the inner cell mass and the outer epithelial trophectoderm [6]. The importance of stage-specific miRNAs throughout the early stages of mammalian embryogenesis is recognized, as is the involvement of coding genes. The first indication of the significance of micro RNAs (miRNAs) in embryonic development was acquired after the ablation of miRNA-treating genes such as Dicer, which is a key player in the start of RNA silencing because it recognizes double-stranded RNAs (dsRNAs) and uses its RNase III-like double-stranded RNA-specific nuclease activity to cleave them into short RNAs and Argonaute RISC Catalytic Component 2 (AGO2), which is a gene that codes for proteins. AGO2 serves a variety of roles, including RNA interference, translation repression, and heterochromatinization, which are all significant in small RNA-guided gene silencing processes [7, 8]. This resulted in an embryonic stop at E6.5 or embryonic mortality at gastrulation. In order to control the quantity or concentration of proteins or enzymes, miRNAs can either degrade mRNA or prevent protein translation. This can either promote or impede embryonic growth. The importance of miRNA before implantation, embryo growth between maternal zygote change, compaction, and blastocyst development has been shown by specific functional investigations or miRNA profiling research. From the zygote and two-cell stage, there is significant breakdown of maternally derived miRNAs, and during mouse embryogenesis, a variety of miRNAs are produced entirely from scratch reduced amounts of intact and hatched blastocysts and elevated Let-7a expression up to the eight-cell stage. The activity of miR-205, miR-150, miR-96, miR-122, miR-146a, miR-145, miR-208, and 496 elevated after bovine minor or major embryonic genome activation. In the initial bovine blast, miR-135a, miR-218, miR-335, and miR-449b are increased [9–12].

These critical developmental processes, such as embryonic cell segregation, embryonic outgrowth growth, and later implantation, are thought to be regulated by miRNA. Throughout these critical stages of embryo development, miRNA dysregulation may have an impact on cell differentiation and subsequent embryo implantation. In this context, Dicer knockout testing revealed that miRNA may retain trophoblast stemness by lowering the activity of Cdkn1a (p21) and Cdkn1c (p57) genes, as well as epiblast pluripotency by avoiding apoptosis. Furthermore, previous research found miRNA in both first trimester and term trophoblast cells, but different miRNAs were expressed in each of these cells, demonstrating that miRNAs may show a part in different trophoblast cell characteristics. The restricted expression of miR-93 on the trophectoderm and upcoming embryonic endoderm, as well as enhanced expression of miR-106a in the inner cell mass, point to a potential involvement for these miRNAs in embryonic differentiation. Additionally, 526 miRNAs, including let-7b, miR-23a, miR-27a, miR-291a, miR-425, and miR-429, showed differential expression among embryos in blastocyst stage and those in embryonic outgrowth stage. Similarly, eight-cell stage embryos and trophoblast tissue have variable expression levels of certain miRNAs [12–16].

miRNAs regulate posttranscriptional gene expression as supported by the fact that around 30% of human genes can have their expression altered by miRNAs, which are distributed in tissue-specific patterns. miRNAs show a significant part in regulating gene expression. Understanding the molecular mechanisms that govern miRNAs is critical since they are frequently dysregulated in a number of human ailments, such as cancer and immunological disorders. RNA polymerase II generally generates major transcripts from miRNA-encoding genes, which are subsequently cut into short RNAs of roughly 21 nucleotides by RNase III endonucleases Drosha and Dicer. The RNA-induced silencing complex (RISC) functions as a posttranscriptional regulator by binding to the 3′-untranslated region (UTR) of mRNAs. It is essential for selecting the start codon and facilitating ribosome binding to the mRNA. Additionally, it significantly influences translational efficiency regulation and the development of the cellular proteome [17], encompassing all miRNAs. This seed-dependent miRNA binding inhibits the translation of its mRNA targets and/or promotes their splicing into nonfunctional forms [18–22]. Unexpectedly, current work provides proof of a target-mediated decay mechanism that regulates the concentration of certain miRNAs. Unexpected miRNA activities have also been found, together with a variety of noncanonical genes harboring miRNA. For instance, some miRNAs work in the nucleus to activate or silence target genes during transcription. These epigenetic alterations are attracted to particular genomic loci by RISC which participates in the breakdown of target mRNA, lowering the amount of transcript accessible for ribosome translation and directed there by miRNAs [23] as shown in **Figure 1**.

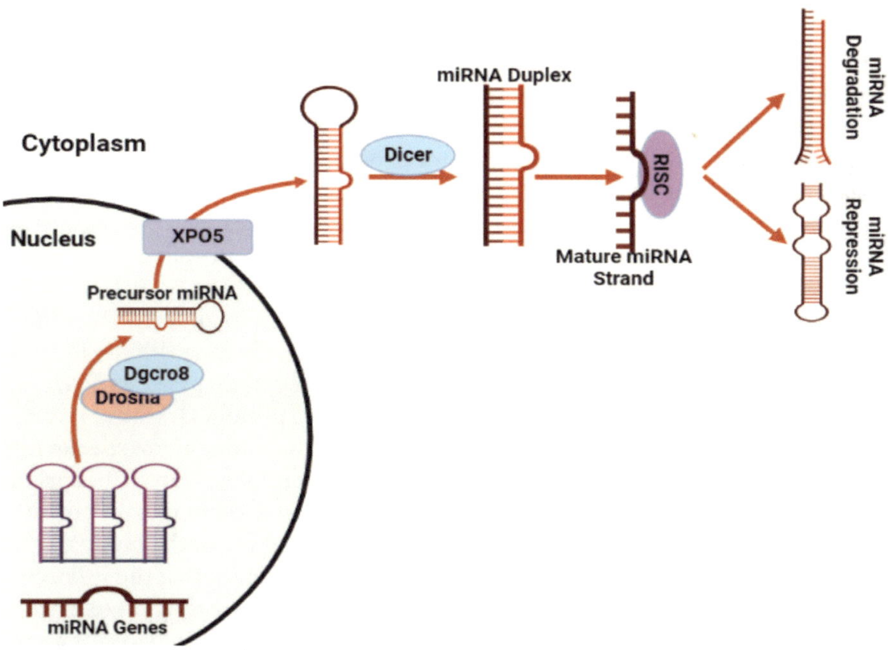

Figure 1.
miRNA synthesis, processing, completion, and control are depicted schematically. The noncoding region of the genome serves as the source for the miRNA transcription within the nucleus. The nuclear membrane protein XPO5 is responsible for the pre-export miRNA's cytoplasmic region from the nucleus. Primary-miRNA is converted into miRNA duplex by the endonuclease dicer in cytoplasm. The mature miRNA is created by when the miRNA duplex unwinds by a helicase. It is the RISC that contains the mature miRNA. Through translation inhibition or mRNA degradation, the miRNA-RISC complex regulates posttranslational gene expression by binding at the mature mRNA.

2. miRNAs biogenesis and mechanism of action

The process of creating miRNAs begins with DNA sequences known as miRNA genes, or with gene clusters that are only transcripted as miRNA molecules or collectively as polycistronic transcripts, correspondingly. A protein-coding gene's intron or non-translated region (UTR) may also be where miRNAs are found. The canonical and noncanonical versions of the two primary biogenesis routes were determined [24].

2.1 Canonical miRNA biogenesis and functions

By use of RNA polymerase II, miRNAs are produced as primary transcripts. Drosha and its cofactor DGCR8 recognize and cleave one or more stem loops that are present in pri-miRNAs. Exportin 5 is a Ran-GTP-dependent mechanism that exports the resultant pre-miRNA from the nucleus. A mature miRNA duplex with two strands that are each 21–22 nt long is produced in the cytoplasm by the processing of the pre-miRNA by Dicer and its cofactor TRBP. After that, the RNA-induced silencing complex (RISC) incorporates this miRNA duplex [25].

To initiate miRNA production, primary miRNA transcripts (pri-miRNAs) from the genome are processed post- or co-transcriptionally. Conventional pri-miRNAs, which are normally transcribed due to RNA polymerase II (POL II), have a 5′-cap, and include one, do not necessarily have a poly-A tail [26].

The majority of mature miRNA genes can express themselves singly or in groups and are found in functional gene coding or noncoding areas. Drosha and Dicer, two RNase III enzymes that catalyze two additional processing stages in the nucleus and cytoplasm, are responsible for the synthesis of miRNAs [27, 28]. In the nucleus, genomic DNA is translated into a long primary miRNA (pri-miRNA). Drosha then cleaves this miRNA to produce a precursor miRNA (pre-miRNA) with a hairpin structure and a length of around 70 nucleotides (nt). These hairpin RNAs are moved nuclear leakage enters the cytoplasm receptor exportin-5, where Dicer cleaves them into mature miRNAs of 19–23 nt active miRNAs can also be produced *via* noncanonical miRNA biogenesis mechanisms. Small nucleolar RNA (snoRNA) progenitors serve as the source of miRNAs. and mirtrons derived from pre-mRNA splicing are examples of these [28, 29].

The microprocessor complex is a protein complex, which involves in the initial steps of processing for microRNA (miRNA) and RNA interference (RNAi) in animal cells [30]. The complex, which converts main miRNA substrates to pre-miRNA in the cell nucleus, is made up mostly of the ribonuclease protein Drosha and the dimeric RNA-binding proteins DGCR8 (also called Pasha in nonhuman species) [31]. Microprocessor also comprises of two units of multiple proteins, transcribes primary miRNAs (pri-miRNAs) from their genes, and then processes them into precursor miRNAs (pre-miRNAs). Introns and noncoding RNA genes are both potential sources of pri-miRNA substrates. In the latter scenario, there is evidence that the spliceosome and microprocessor complex interact and that pri-miRNA processing takes place before splicing [32]. Its two components have been found to be both necessary and sufficient to mediate the formation of miRNAs from pri-miRNAs [30]. One of them is a massive multiprotein complex made up of multiple RNA-associated proteins, ribonucleoproteins, dsRNA-binding proteins, and the family of Ewing's sarcoma proteins. The other is Drosha, tiny multiprotein complex that contains the enzyme RNase III, as well as the DiGeorge syndrome critical region 8 RNA protein complex (DGCR8). The microcontroller identifies certain pri-miRNA patterns and splits it at the bottom of the distinctive hairpin form, resulting in a pre-miRNA, there is a 3′–2–nt overhang and a 5′–monophosphate. Afterward, a complex of exportin5 (XPO5) and the nuclear protein guanosine-5'-triphosphate-ase, which is connected to RAS, was formed (Ran-GTPase). This complex exports pre-miRNAs from the nucleus to the cytoplasm. The RNase III endonuclease Dicer eliminates the terminal loop of pre-miRNA in the cytoplasm [33, 34]. miRNAs are currently loaded, entering the complex for RNA-induced silencing (RISC) as short RNA duplexes called miRNA duplexes [34, 35].

2.2 NonCanonical miRNAs

There are miRNAs known as noncanonical miRNAs that are produced using a distinct biogenesis mechanism. Recently, different miRNA biogenesis routes have been identified. Certain noncanonical pri-miRNAs, for example, are known as mirtrons and are stored in coding genes' introns [23]. Like typical introns, all mirtrons undergo initial processing by the nuclear splicing machinery before creating sturdy, stem-less hairpins than canonical pri-miRNAs. Only one annotated miRNA has been found to process independently of DICER so far. Pre-miR-451 miRNA maturation necessitates the AGO2 slicer activity because of its stem-loop shape is too shorter for DICER to cleave [23]. Pre-miR-451 processing by AGO2-mediated processes and eventual RISC loading are influenced by the length of the stem-loop structure, inappropriate base pairing in the stem, and low GC content in the distal stem, among other factors. Furthermore, miR-451 processing by AGO2 is dependent on the function of eukaryotic translation initiation factor 1A, another RISC component (EIF1A) [23].

It has been predicted that mirtron formation occurs during noncanonical miRNA biogenesis. RNA Pol II transcribes mirtrons as introns of protein-coding genes. These mirtrons are spliced by the spliceosome, debranched, and (if necessary) exosomal trimmed. Mirtrons are considered as canonical miRNAs after undergoing these preliminary processing stages, and Exp5 then transports them to the cytoplasm, where they are processed by Dicer and TRBP and then integrated into the RISC. Also, it has been highlighted the production of miRNAs independent of Dicer. RNA Pol II produces the Dicer-independent miRNA miR-451, which is then processed by Drosha/DGCR8 and exported from the nucleus by Exp5. Argonaute 2 (Ago 2) binds to miR-451; once it is inside the cytoplasm, where it is then cut into the functional, mature, single-stranded miRNA [25].

2.3 shRNA- snoRNA- and tRNA-derived miRNAs

The family of ncRNAs known as snoRNAs, which range in length from 60 to 300 nt, is typically found in cells' nucleolus, where they may alter tRNAs rRNAs and snRNAs. Through the H/ACA box on their rRNA precursors, snoRNAs can control the methylation and pseudouridylation of specific rRNA sites. According to a number of lines of evidence, certain snoRNAs can serve as the building blocks for noncanonical miRNAs. These noncanonical miRNAs are produced by use of a biological process, which resembles the synthesis of canonical miRNAs in that it involves the Dicer, Dgcr8, and AGO enzymes. Dicer and Dgcr8 are required for the stabilization and degradation of the purportedly derives from snoRNA miRNAs [36]. SnoRNA-derived miRNAs are 21 nt long, equivalent to miRNAs, and can link to AGO enzymes such as AGO1 and AGO4 to inhibit mRNA targeting [36, 37]. Contrarily, the production of two snoRNA-derived miRNAs called ACA45 and GlsR17 shows a dependence on Drosha/Dgcr8 rather than Dicer [36, 37]. The unusual precursor structure of these snoRNA-derived miRNAs consists of two hairpins resembling pre-miRNA connected *via* a hinge. These distinct assemblies appear to give their capacity to silence mRNA transcripts posttranscriptionally, similar to miRNAs. ShRNA-derived miRNAs, such as snoRNA-derived miRNAs, are derived from non-miRtronic genomic areas and require Dicer but not Dgcr8 for synthesis. This subclass includes the isoleucine tRNA gene, shRNA-miR-320, miR-484, and miR-1980. They are all distinguished by distinctive physical characteristics, such as structures with hairpins that surround the conventional pre-miRNA hairpin but are invisible to Dgcr8. These shRNA-derived miRNAs have a complicated and unusual hairpin structure that is located at the 3′ end rather than the traditional 5′ end. Emerging information on shRNA-derived miRNA clusters suggests that a new biogenesis may represent additional crucial regulatory activities that have not yet been recognized [23, 36, 37].

tRNA-derived miRNAs are created through tRNA development, although they have activities comparable to miRNAs, such as RNA transcript suppression. However, compared to traditional miRNAs, their affinity for Ago proteins appears to be less., which is probably because of their susceptibility to posttranscriptional modifications. Although little is known about their function, tRNA-derived miRNAs may support cell growth and miRNA biogenesis [37–40].

Exportin 5 transports a pre-miRNA hairpin enters the cytoplasm and forms a miRNA duplex is formed following a Dicer's second cycle of endo nucleolytic cleavage [41]. Certain miRNAs are processed by Dicer a few organisms through interactions with supplementary RNA-binding proteins such as loquacious and TRBP. [42]. These RNA-binding proteins' posttranslational modifications can influence Dicer activity and the biogenesis of specific miRNAs [29]. Following duplex unwinding and strand selection, the mature miRNA is loaded into the Argonaute protein, which the miRNA-induced suppressing complex's core component (miRISC) [41]. Through sequence complementarity, the targeting mRNAs' 3'UTRs are bound by the miRISC-associated miRNA, which also attracts downstream effectors and typically causes translational suppression and mRNA degradation [43]. The seed sequence is the first 2–8 nucleotides from the mature miRNA's 5′ end. To pair with a target mRNA, the seed sequence must be nearly full, while exhibiting partial symmetry along the remainder of the miRNA and the target. MiRNAs that have the identical seed pattern are members with the same family of miRNAs. A plethora of computational and biological approaches have been developed to anticipate and confirm miRNA-mRNA target pairings [44].

Argonaute posttranslational changes can influence steady-state miRNA activity [45]. The orthologs of Argonaute in humans and *C. elegans*, AGO2 and ALG-1, have conserved phosphorylation sites that modulate mRNA target binding, according to mass spectrometry [46, 47]. To continue the overall a phosphorylation cycle, the effectiveness of miRNA-target associations in different cell lines controlled by the casein kinase CSNK1A1 and the phosphatase ANKRD52-PPP6A suggested, in which specific binding causes Quick phosphorylation and dephosphorylation of AGO2 [48]. Phosphorylation of the Argonautes AGO1 and AGO2 proteins was discovered to be essential for interaction with LIM domain-containing proteins in a separate study. To promote target silencing by certain miRNAs, these LIMD proteins establish a clamp between Argonaute and the adaptor protein GW182. In the absence of that phosphorylation event, AGO3, which has a phosphomimetic residue near the phosphorylation site of AGO1 and AGO2, is recruited to execute the identical silencing effect [49]. These findings imply that the Argonaute function may be situation sensitive. The production of a shortened version of AGO2, deficient in every functional area, has been proposed as the cause in mouse oocytes of a general downregulation of miRNA activity. In the absence of mature miRNAs, the E3 ubiquitin ligase Iruka poly-ubiquitinates AGO1 at an unaltered lysine position, designating AGO1 for destruction [50].

3. miRNAs in early embryonic development

3.1 The role of miRNAs in preimplantation development

The free-living stage of mammalian development from the one-cell zygote to blastocyst adhesion or implantation into the uterine wall is known as preimplantation development. The first week of development is typically covered by it, and key moments such the change from parental to embryonic genomic activation (EGA), compaction, and cavitation or blastocyst formation are included [51]. The primary result of preimplantation development is the creation of an embryo that is implantation-competent for the start of pregnancy. To achieve this, the trophectoderm (TE), which develops into the blastocyst's outer layer and connects with the endometrium to start implantation, must first take shape. The embryonic element of the placenta is likewise derived from the TE. The initial stage of the embryo proper is the inner cell mass (ICM), a second cell type that is still pluripotent. These cells continue to be contained within the polar and mural TE of the blastocyst. Therefore, TE differentiation rather than embryogenesis is the focus of preimplantation development as placental mammals must first meet this criterion for implantation and placental growth before embryogenesis can occur [51].

The animal embryo moves between the oviduct and the uterus after fertilization. The embryo then implants, forming a link with maternal tissues as depicted in **Figure 2**. Embryonic development in mammals a diploid zygote is created when a sperm and an egg are united. The zygote proceeds through a sequence of cleavage divisions that increase the number of blastomere-containing cells while maintaining the embryo's total volume. Following cleavage, the eutherian embryo goes through compaction, in which the compaction of the embryo one another to form a tight ball of cells known as a morula [52].

miRNAs, which are around 22 nucleotides long, provide guidance molecules in the RNA silencing process. MiRNAs have a role in nearly every animal developmental and pathological process, targeting the majority of protein-coding transcripts [53]. A crucial step is embryo implantation. In the creation of a pregnancy in animals and happens over a "period of implantation" is a certain time frame [54]. Apposition, attachment, and invasion are the three stages of implantation [55]. The receptive endometrium, which is produced by widespread endothelial edema on the surface of the inner uterus, interlaces with the microvilli of the cytotrophoblast on the outer surface during apposition [56]. Furthermore, the receptive endometrium expresses heparin-binding growth factor similar to EGF (HB-EGF), and cells expressing Blastocysts are adhered to by transmembrane HB-EGF with ErbB4 on their cell surface [57]. Activated blastocysts increase the expression of HB-EGF, which causes its own expression of genes at the blastocyst apposition location in the endometrium *via* an auto-induction loop [58]. Lif regulates HB-EGF expression, which leads to a decrease in COX-2, a deficit of which results in implantation failure [59]. At the location of implantation, gene expression in the Wnt/−catenin pathway is also crucial. Wnt/−catenin signaling prior to attachment necessitates an active blastocyst and preimplantation. Transgenic mouse models of estrogen secretion [60].

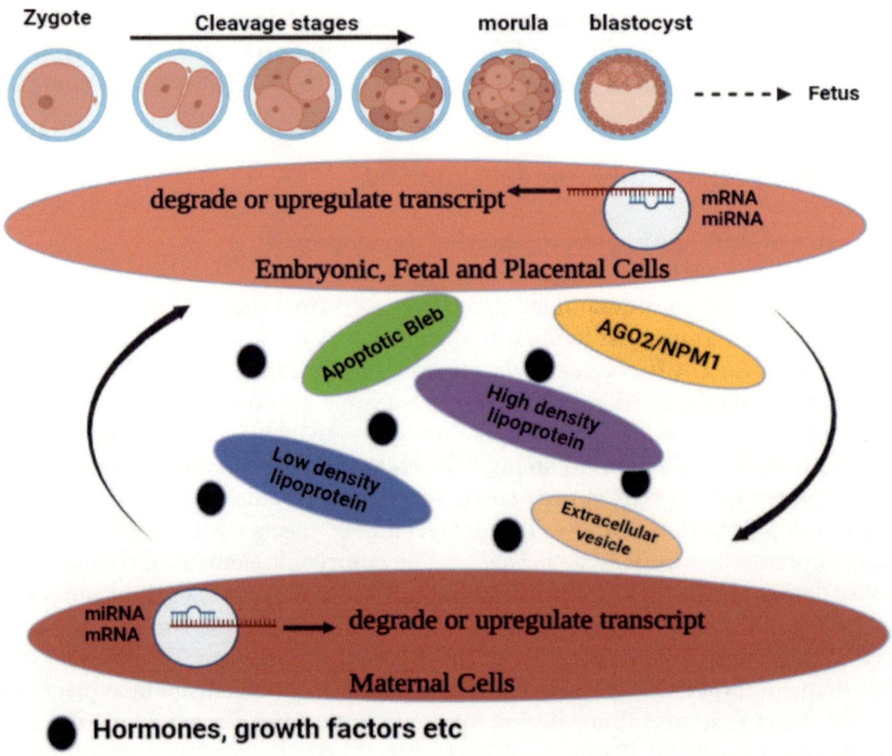

Figure 2.
miRNA transfer between the mother and fetus is depicted schematically. miRNAs may be transported in between fetal cells and the conceptus using a variety of carriers. Proteins such as extracellular vesicles, cytotoxic benign growths, and high-density and low-density lipoproteins, such as Argonaute 2 (AGO2) and Nucleophosmin 1 (NPM1), can all act as these vehicles. The miRNA pairs with a complementary target mRNA when it is delivered to the mother or conceptus and joins an RNA-induced silencing complex (RISC). The desire mRNA is subsequently dysregulated, increased, or degraded. Direct cell-cell contact can cause this crosstalk, or miRNAs can move around and be conveyed through fluids, such as uterine fluid.

During the preimplantation stage of development, the embryo secretes extracellular miRNAs are introduced. MiRNAs are secreted into biological fluids in the extracellular setting by a variety of cell types. These miRNAs generated from fluid have been demonstrated to circulate in the body. The appropriate packing of vesicles that contain miRNAs outside of cells, especially exosomes, is required for stable transport. At the forefront of studies into cell-to-cell communication are these vesicles, which transport RNA, DNA, and proteins. The presence of EVs in their miRNA content was found in uterine fluid. Their miRNA content was found in uterine fluid, which is expected to facilitate interaction between the mother and conceptus. Uterine fluid is one of the reproductive fluids where they have been discovered (**Table 1**) [61].

3.2 The importance of miRNAs in the establishment of the three germ layers

A large a group of noncoding RNAs known as miRNAs function as RNA silencing guide molecules and are approximately 22 nucleotides long. Animal processes, both developmental and degenerative virtually, universally involve miRNAs, which primarily target protein-coding transcripts [53]. The “window of implantation” refers to the constrained time period during which embryo implantation, an essential

miRNAs	Function	References
miR-143	Proliferating, migrating, and invading cells of human endometrial stroma	[62]
miR-193	In the uterus of mice, implantation rate is controlled.	[63]
miR-181	Control of the implantation rate in the mouse uterus	[64]
miR-29b	Development of a calf using somatic cell nuclear transfer (SCNT)	[65]
miR-126a-3p	Development utilizing nuclear somatic cell sojourning in bovine embryos (SCNT)	[66]
miR-200a	Counting the implantation spots in the rat endometrium	[67]

Table 1.
Role of miRNAs in the development and implantation of embryo.

step in the creation of pregnancy in mammals, takes place (WOI) [54]. Apposition, attachment, and invasion are the three stages of implantation [55]. The receptive endometrium, which is produced by widespread an inner uterine surface with stromal edema, interlaces with the microvilli of the cytotrophoblast on the outer surface during apposition [56]. Additionally, the receptive endometrium expresses heparin-binding growth hormone similar to EGF (HB-EGF), the ErbB4-expressing cells that participate in this adhesion to blastocysts [57], and the ErbB4-expressing cells that participate in this adhesion to blastocysts *via* an auto-induction loop [58]. Lif regulates HB-EGF expression, which leads to a decrease in COX-2, a deficit of which leads to unsuccessful implantation [59]. At the location of implantation, which results in the failure of the implant. Wnt/-catenin signaling prior to attachment necessitates an active blastocyst and preimplantation. In transgenic mouse models of estrogen secretion [60] various glycoproteins, ligands for carbohydrates, receptors, and integrins collaborate the uterine surface in order to secure the embryo during the attachment phase. Several integrins play a role in implantation in varying degrees. The embryo expresses 51, v3, v5, and v6 during attachment, and 111, 61, and 71 are later implicated in invasion [68].

The uterine surface in order to secure the embryo into biological fluids in the extracellular environment by a variety of cell types. These miRNAs generated from fluid have been demonstrated to circulate in the body. The appropriate packing of extracellular vesicles containing miRNAs (EVs), especially exosomes, is required for stable transport. The study of cell-to-cell communication is currently focused on these vesicles, which transport RNA, DNA, and proteins. EVs have been discovered in a number of reproductive fluids, such as uterine fluid, where it is believed that their miRNA content serves as a mechanism for mother and child chit-chatting conceptus [61].

3.3 The role of miRNAs in cell proliferation, differentiation, and apoptosis during embryogenesis

MiRNAs have a significant impact on a variety of cellular processes, such as differentiation and proliferation, that are regulated by intracellular signaling. Major examples of these signaling cascades are the nuclear factor kappa-light chain-enhancer of activated B cells (NFkB), mitogen-activated protein kinase (MAPK), phosphatidylinositol 3-kinases (PI3K), notch, transforming growth factor (TGF), and hedgehog (Hh) pathways. MicroRNAs may regulate cell signaling by suppressing the expression of their target genes, which produce molecules involved in these important signaling pathways [69, 70]. Additionally, a number of signaling mechanisms control miRNA expression. For instance, in rat vascular smooth muscle cells (VSMC), an active MAPK pathway increases miR-31 expression. In PC12 cells, a rat pheochromocytoma cell line of the adrenal medulla, the nerve growth factor (NGF) increases expression of miR221, miR-222, with NGF induction itself dependent upon activation of the extracellular signal-regulated kinase 1 and 2 (ERK1/2) pathways [71]. In signaling cascades, microRNAs may potentially establish regulatory feedback loops. In many physiological processes, there are several regulatory loops between miRNAs and their targets. In THP-1 cells, a cell line for human acute monocytic leukemia, a regulatory loop between miR-146 and NF-B has been identified [72]. Through mediating inter-pathway crosstalk, microRNAs may operate as a connecting agent between several signaling pathways [70, 73]. For instance, in the early stages of embryonic patterning, miR-15 and miR-16 link the Wnt and TGF- signaling pathways. To sum up, miRNAs play a role in controlling intracellular signaling pathways. Additionally, they create intricate links between various signaling pathways. Changes in intracellular pathways are linked to a number of diseases in humans.

Preimplantation embryos from mice and humans can produce the placenta, yolk sac, and the three germ layers of the fetus, respectively, through the actions of the epiblast (EPI), primitive endoderm (PrE), and trophectoderm (TE) [74]. Years of work have resulted ESCs, XEN (extraembryonic endoderm) stem cells, and trophoblast stem cells are three *in vitro* models for cell culturing reproduce the potentials of those three distinct lineages (TSCs) [75]. Shortly, after implantation, it is possible to identify two additional stem cell types: two stem cells: the primordial germ cell (PGC), the very first progenitor of the germ cell lineage, and the stem cells from epiblasts (EpiSC), which mimics the potential of the more primed epiblast [76]. It is worth noting that the expression patterns of the three preimplantation embryo lineages varied significantly between mice and humans [77].

MiRNAs can influence adult stem cell destiny in a variety of ways; a type of stem cell called epiblast (EpiSC), and hematopoietic stem cells (HSCs) hinders whereas miR-16, miR-103, and miR-107 expression decrease proliferation, and they promote differentiation [78]. MiRNAs have been shown to influence a variety of differentiation processes such as myogenesis, cardiogenesis, neurogenesis, and osteogenesis [79].

miRNAs linked with stem cell development are expressed is regulated at several levels, for instance, posttranscriptional, epigenetic, and/or transcriptional control; cluster of miR-362 and miR-367, for instance, is regulated by transcription factors (TFs) connected to the ESC, including Oct3/4, Sox2, and Rex1 [80], which promote the self-renewal or generation of pluripotent stem cells. Studies on the transcriptional control of miRNA production in ESCs have shown, which bicistronic or monocistronic miRNAs at intergenic loci are regulated by shared upstream transcription regulators, bicistronic or monocistronic in nature. Along with the bi-cistronic miRNAs mir-1 and mir-133, mir-206, mir-486, and mir-499, are induced by MyoD, Myogenin, and MYF5 during human muscle stem cell (hMSC) growth [81]. It is interesting to note that the co-regulated miRNAs mir-1 and mir-206 in various Chromosomes share a seed sequence with members of the same miRNA family that downregulates relevant targets [82].

Cells that would typically be harmful to the organism's survival are eliminated during apoptosis, a sort of active, and cell-autonomous caused cell death.

The so-called intrinsic pathway, which includes mitochondria and stressors, can start this death response. Alternatively, the extrinsic system, which involves a cognate ligand, might start it by activating death receptors on the cell membrane [83]. In 2003, the first papers on the regulatory functions of miRNAs in apoptosis were published. Through the inhibition of the effector caspase Dicer and the proapoptotic high temperature-induced Dauer synthesis family member (hid1) genes, respectively; it was demonstrated in these Drosophila research that miR-14 and bantam govern cell death. Since then, both intrinsic and extrinsic routes have been revealed to contain many miRNA families that regulate apoptosis. The discovered miRNAs have possibly pro- or antiapoptotic characteristics [84].

Extrinsic signaling pathways are occupied by a large number of miRNAs. miR-7 specifically sensitizes cells from hepatocellular cancer and glioblastoma multiforme to TRAIL-mediated apoptosis [85]. In human acute myelogenous leukemia cells to improve cell survival, miR-590 targets Fas ligand (FasL) and is regulated utilizing the STAT5 (Signal Transducer and Activator of Transcription) pathway [86]. MiR-21, a direct FasL target in pancreatic cancer, prevents cancer cells from dying from gemcitabine-induced apoptosis when it is expressed ectopically [87]. Fas expression in osteosarcoma cells is decreased by miR-20a, promoting tumor cell proliferation, and metastatic potential [88]. Additionally, miR-146a and miR-196b prevent the expression of Fas, and leukemia is brought on by ectopic miR-196b expression [89]. The expression of Fas is also inhibited by miR-146a and miR-196b, and leukemia is caused by ectopic miR-196b expression [89]. miR-25, which is associated with intrahepatic bile duct. It is more common for malignancy to directly target DR4 and resist TRAIL-mediated apoptosis [80]. There is an increase in malignancy, which has been demonstrated to directly target DR4 and shield against TRAIL-mediated apoptosis. Although miR-128a antagonists increase Fas-mediated apoptosis, ectopic miR-128a In Jurkat cells, expression raises Fas resistance by targeting FADD [90]. By inhibiting the expression of miRNA-186*, which targets the caspase-10 enzyme, which is identical to caspase-8, curcumin induces death in A549 cells [91]. MiR-708 improves the susceptibility a kidney cancer cells multiple apoptotic stimuli and adversely controls the expression of c-FLIP [19]. Three miRNAs, miR-378, miR-155, and miR-let-7a, target caspase-3 that can regulate the apoptosis of cancer cells [92]. Specifically targeting caspase-7, miR-106b is linked to the prostate cancer coming back [93]. For the purpose of controlling pancreatic cancer cells undergo apoptosis,miR-133 and miR-24a directly inhibit caspase-9 [94].

A variety of miRNAs are also involved in the regulation of the apoptotic regulator genes in the intrinsic route for apoptosis. It has been demonstrated that miR-365 specifically targets the protein adaptor Src homology 2-containing subtype 1and the Bcl-2-like protein 4 (Bax) (SHC1)in lung adenocarcinoma and lung cancer cell lines [95]. In gastric cancer cells, miR-125b expression is markedly downregulated. Overexpressing miR-125b, however, prevents stomach cancer cell lines from proliferating, migrating, and invading (HGC-27 and MGC-803) (95)miR-148a and miR-204 cause apoptosis *via* posttranscriptionally targeting Bcl-2 negatively [96]. The posttranscriptional level of miR-148a and miR-204 negatively targets Bcl-2 to cause apoptosis [97]. By triggering apoptosis, miR-608 greatly reduces the growth of chordoma cells by directly targeting Bcl-extra-large (Bcl-xL) [98]. Activator of apoptosis protein 1 (Apaf1)can be targeted by miR-23a/b and miR-27a/b, which controls the vulnerability of neurons to apoptosis (**Table 2**) [99].

miRNA	Function	References
miR-21-3p	Inhibits the cow granulosa cells undergo autophagy	[100]
miR-383	Prevents autophagy in granulosa cells of cattle	[101]
miR-183-96-182 cluster	Controls apoptosis in the granulosa cells of the porcine ovary	[102]
miR-1275	Involved in the synthesis of estradiol and death in porcine granulosa cells	[103]
miR-92a	Controls the apoptosis of granulosa cells in porcine ovary	[104]
miR-15a	Controls the granulosa cells in the human ovary produce steroid hormones and promote cell growth	[105]
miR-96	Controls the growth of cells and the synthesis of steroid hormones in human ovarian granulosa cells	[106]
miR-126-3p	Encouraging in porcine granulosa cells, cell growth, and averting cell death	[107]
miR-181a	Control oxidative stress-related acylation of the fork head box protein O1 (FOXO1) and mouse granulosa cell apoptosis	[108]
miR-23a. miR-27a	Granulosa cells in humans undergo apoptosis	[109]
miR-99a	Polycystic ovarian syndrome (PCOS) sufferers' granulosa cell growth and death are controlled.	[110]
miR-335-5p	In women with polycystic ovarian syndrome (PCOS), the regulation of granulosa cell growth and apoptosis is in place.	[111]
miR-125b	Controls the apoptosis of the yak ovary's granulosa cells	[112]
miR-143	Encourages human granulosa cell proliferation and estradiol synthesis that is produced by follicle-stimulating hormone (FSH)	[113]

Table 2.
miRNAs that are involved in autophagy, cell proliferation, and apoptosis.

4. miRNAs in organogenesis

4.1 The role of miRNAs in organogenesis and tissue differentiation

miRNAs may be crucial for differentiation and organogenesis in *D. rerio*, according to research on Dicer mutant animals. Maternal-zygotic Dicer mutants were discovered to have disruptions in several developmental processes, including neurogenesis, somatogenesis, and the development of the ear, eye, and heart [114]. It is not yet known if every one of these flaws results from a deficiency in miRNAs or from other Dicer-dependent mechanisms. There are currently no zebrafish strains with miRNA knockouts; however, rescue experiments on maternal-zygotic Dicer mutant animals show that the miR-430 family of miRNAs plays a crucial role in neurogenesis [114]. Defects in neurulation are severe in maternal-zygotic mutated Dicer embryos. These mutants have problems forming the neurocoele and neural tube, as well as the midbrain-hindbrain boundary and brain ventricles. Surprisingly, this mutant phenotype was partially reversed by miR-430, and a single miRNA was injected, which belongs to a broad kinship with it belongs to the miR-302 and miR-17 families of miRNAs . The midbrain-hindbrain border was saved after miR-430 injection but not after another unrelated miRNA. Additionally, the injection of, however, heart or ear development was not partially rescued by miR-430, only somatogenesis, retinal development, and gastrulation. These findings demonstrate conclusively that the loss of miRNAs is the cause of at least some of the aberrant phenotypes seen during maternal-zygotic diverse mutant animals. Without a miRNA knockdown strain, a unique system is provided by the maternal-zygotic Dicer knockout to study the function of additional miRNAs in zebrafish.

M. musculus is comparable to *D. rerio* and comes with the same warnings. MiRNAs play a variety of mouse embryogenesis, according to Dicer1 mutant mice, which plays roles in differentiation and development [7]. Conditional-knockout strains in the mouse can be used to explore the necessity of both Dicer and miRNAs in various tissues at various developmental junctures, providing further insights. Conditional gene targeting led to the creation of ES cells with the Dicer mutation, which exhibit severe differentiation abnormalities *in vitro* [115]. MiRNAs appear to operate in cells, lineage commitment, not however in CD4/CD8; according one study found that examined the consequences early in T-cell development by genetically inactivating Dicer1 [116]. MiRNAs seem to function in various cellular processes and lineage commitment, but interestingly, they do not seem to play a role in CD4/CD8 development. A study investigating the effects of genetically inactivating Dicer1 during early T-cell development supports this observation [117].

This conditional-knockout research in mice has a special warning because it is frequently difficult to determine how effectively any and the Dicer preexisting When Dicer1 is deleted somatically, miRNA pools are depleted. In reality, miRNAs do seem to last a while [116]. One particular miRNA, MiR-181, has been directly associated to the maturation of B cells [118]. This miRNA is highly expressed in mouse B-lymphoid cells. Hematopoietic stem cells overexpress some genes and increases the proportion of B-lineage cells both *in vitro* and *in vivo* in adult mice during differentiation tests. Dicer1 was conditionally inactivated in particular areas of the mouse limb mesoderm, which resulted in, however, there are no mistakes in fundamental limb patterning or tissue-specific differentiation, only mutant embryos exhibit severe growth defects limbs [119]. This is an astonishing discovery that is similar to the Dicer1 victory over *D. rerio*. The Dicer1 victory over *D. rerio*. It is still unclear, though, whether lingering deciphering or miRNA pools may have covered up prior miRNAs' functions in the growth of limbs.

The Dicer1 knockdown in *D. rerio* has some similarities to this startling discovery. If any residual deciphering or miRNA pools may have covered up earlier miRNA the development of limbs, this is still unknown. [120]. Myocardin, Mef2, serum response factor (Srf), and Myod have all been found to operate as transcriptional controllers of *in vitro* miR-1 expression. Through investigation of the upstream-regulatory sequence of these two genes [121]. By employing a conditional Srf-knockout strain, it was discovered that Srf is necessary the expression of miR-1in a mouse heart that is still growing [116]. Developmental arrest occurred around heart failure caused by upregulation of miR-1 under the promoter of the -myosin heavy chain on embryonic day 13.5. ventricular cardiomyocyte proliferation anomalies and thin ventricle walls appeared in transgenic embryos. Mice with a mutated form of the miR-1 gene exhibited reduced levels of the transcription factor Hand2, even though there was no apparent alteration in Hand2 mRNA levels. This implies that Hand2, a potential miR-1 target in myocardial development, is affected [121].

5. The importance of miRNAs in the formation of specific organs

5.1 Heart

Each of the several stages of development involves a number of substances with inducing or suppressing capabilities [122].

Preliminary cardiac precursor cells have been found in the mouse and chick embryo's front portion of the primitive streak, according to tracing investigations using studies using cell/tissue explants and essential dyes [123]. The FGF family of molecules' action signaling molecules, which includes the FGFs FGF2, FGF4, and FGF8 have been postulated to indicator through FGFR1, controls the initial movement of these cells from the posterior to the anterior at the start of gastrulation [124, 125]. Similar to TGF-, bone-forming protein 2 (BMP2), a growth factor belonging to the TGF- superfamily, is first expressed toward the final phase of the primitive phase by the start of gastrulation and is associated with early cell migrations [126].

As a result of Mesp-1 and Mesp-2 being expressed in an FGF-8-dependent manner at embryonic day 6.5 (E6.5); Bilaterally, cardiac progenitor cells move to the front and side of the heart of the embryo, colonizing the lateral plate mesoderm splanchnic layer, and forming two early lateral cardiac fields [127]. Mesp-1 positive mesodermal lineage is induced by canonical Wnt/–catenin signaling in the neural ectoderm during early gastrulation [128]. These findings support significant roles for early Bmp2 and FGF signaling in establishing cardiogenic potential [129], limiting the position of the cardiac fields, and directing progenitor cell movement. By acting as a concentration gradient of Wnt/–catenin, the underneath anterior mesoderm can be induced into head tissue [130].

A day later (E7.5), the cardiac crescent, a horseshoe-shaped epithelial fold, is formed when these earliest cardiogenic stem cells travel cranially and congregate across the midline region of the embryo. Nkx2-5, a vertebrate homeodomain transcription factor and *D. melanogaster* tinman gene homolog, was first used to identify the lateral cardiac fields in mouse and chicken embryos [131, 132]. Nkx2-5 is the Nkx factor that is most abundantly expressed during early peripheral cardiac fields and the cardiac crescent undergoes cardiogenesis in mouse and chick embryos [133, 134]. But unlike the way Tinman works in flies, no Nkx gene controls cardiogenesis in a master manner [135]. The lateral portions of the cardiac crescent are subsequently drawn toward the midline by a dynamic process that causes the medial area of the crescent to expand and travel cranially; these two lateral cell populations eventually unite to form the heart tube (E8.0). The first heart areas can finally be anatomically and molecularly recognized when the heart tube is forming. Wnt/–catenin signaling, particularly Wnt-3a and Wnt-8, is required for the initiation and upkeep of Isl-1 expression during the development of the heart tube [136] and Fgf10 [137]. The development and stability of Isl-1 expression during the development of the heart tube depend on Wnt/–catenin signaling, particularly Wnt-3a and Wnt-8 [138].

At E9.5, the cardiac tube is bent to the right to adjust the heart's chambers (looping), and the caudal region of the cardiac tube is brought cranially to position the two sinus venosus horns dorsally to the outflow tract. Since the sinus venosus moved the right side of the heart by E10.5, the cardiac looping process is almost finished. The right atrium and common cardinal vein on the right are joined at this point, which is indicated by the emergence of two outgrowing structures known as the sinus valves. Additionally, at E10.5, the cardiac chambers have begun to septate [139, 140]. By E11.5, the heart's caudal region is at last dorsal to the arterial pole, and in oxygen-breathing vertebrate species, the heart chambers will then become septate is the last step in their specification. The outflow route is initially divided into an arterial and pulmonary arch by the creation of internal cushions. The left horn shrinks and changes into the coronary artery as the right sinus venosus horn is placed in close proximity to the right atrium. The interatrial and interventricular septa, respectively, divide the atria and ventricles [125, 141].

5.2 Lung

Similar divisions between the stages of lung development in mice and humans, including canalicular, terminal saccular, and pseudoglandular, and the final alveolar stage, are present in both species [142]. Total miRNA levels in the mouse embryo at the two-cell stage were found to be considerably lower than those in the one-cell zygote following fertilization [9]. A noteworthy difference between the two-cell stage and the four-cell stage embryos was the overall miRNA levels [9]. Inhibiting sections of cultured tissue from mouse lung E12 smoothened led to an increase in the distal epithelial size, as well as to disturb the usual branching pattern and mesenchymal integrity [143]. Inhibiting smoothened in cultured explanted E12 mouse lung caused distal epithelium to enlarge and disrupted the mesenchymal integrity and normal branching pattern. Mesenchymal progenitors' growth and differentiation in the lung development of mice were controlled by miR-142-3p after it was all said and done [144].

The amount in the forming mouse lung miR-17 was steady from E11.5 to E16.5; it was mostly present at the pseudoglandular stage, with greater epithelial levels around E12.5 [145]. MiR-17 levels in the developing mouse lung were steady (E11.5–E16.5), primarily when pseudoglandular with greater epithelial levels at E12.5 [145]. In the rat embryonic lung, miR-449a levels increased from E15.5 to E18.5, and miR-449a was upregulated in the human fetal lung at 18–20 weeks (canalicular stage) [146]. Levels of Mycn and Sox9 mRNA and SOX9 and Ki-67protein were all elevated at the distal end of the epithelium of E16.5 mouse lung culture when miR-449a was inhibited (end of pseudoglandular stage) [146].

The deletion of miR-26a-1 and miR-26a-2 in mice enhanced the creation of dilated lumens and aerated patches at the beginning of the neovascularization stage (E16.5 of lung development) [147]. The miR-26a mutant mice's results at the level of secularity (E18.5) likewise showed enlarged crystalline bodies and higher amounts of the proteins SP-A, SP-B, and SP-C. The findings also suggested that miR-26a is implicated in the pulmonary surfactant generation [147]. MiR-17-92 cluster and miRNAs in the Dlk1–Dio3 locus were upregulated in the E18.5 mouse lung after epithelial histone deacetylase 3 was lost [148]. Embryonic mouse lung epithelium overexpressing miR-17-92 indicated densely packed AT1 cells. The results of the study collectively demonstrated that histone deacetylase 3 controlled miR-17-92 levels, which, in turn, affected TGF- signaling and, ultimately, AT1 cell spreading and saccular formation [148]. The highest amount of miR-127 was found in the fetal rat lung at E21 [149]. Additionally, the size of the terminal and internal buds was increased by overexpressing miR-127 in fetal rat lung (E14) culture.

In research on alveolarization, bronchopulmonary dysplasia (BPD) mouse models have been used. After, the treatment of miR-876-3p mimics repaired alveolar architecture and reduced neutrophilic irritation in postnatal day 14 animals that had been subjected to hyperoxia [150]. After exposure to hyperoxia, levels of miR-30a-3p and -5p were significantly increased in the lungs of female mice on postnatal days 7 and 21 [151]. In total, significant reductions in miR-30a-3p and -5p levels were observed, whereas DLL4 mRNA and protein levels were elevated in human BPD lung samples [151].

5.3 Brain

A KO would be very difficult because of the 12 different members of the miRNA Let-7 family, which is strongly expressed in the brain. In addition, miRNAs are frequently transcribed in groups. Therefore, the manifestation of a different unrelated miRNA in the same cluster may be affected if one miRNA gene is deleted or otherwise altered. Along with having overlapping targets, miRNA families also have the same seed sequence. It follows that other family members would probably make up for the loss an individual's family. Even with these challenges, reports of miRNA knockout (KO) mice have allowed for loss-of-function studies for particular During developing brain, miRNAs [151].

One of the most abundantly produced miRNAs in the mouse brain is miR-124, which is also highly conserved among mammals. When neural stem cells differentiate into neurons, miR-124 expression rises and is neural-specific [152, 153]. Reduced neurogenesis and a shift to gliogenesis are caused by neural stem cells permanently losing the ability to function as miR-124 [154]. MiR-124 is a neuronal destiny determinant, as determined by the forced neural differentiation that occurs when it is overexpressed in glioma cells, embryonic stem cells, HeLa cells, and neural progenitor cells [154–157]. Three distinct loci for the miRNAs mIR-124-1, mIR-124-2, and mIR-124-3 yield the same mature sequence and are used to transcribe miR-124. Retinal noncoding RNA gene 3 has an exon from which the miR-124-1 stem-loop is translated (Rncr3). The nuclear localization, consensus sequence, and high amounts of miR-124 expression in the brain are all due to this gene, which also serves as its main source. In order to study the function of miR-124-1 during brain development, Rncr3 was deleted to create miR-124-1 knockout (KO) mice. In this work, it was discovered indicated there was a decrease in miR-124 expression by 60–80% in KO mice. Around postnatal day 20, the majority (60%) of KO mice perished. In addition to a large loss of cone cells and cell apoptosis in the cortex and hippocampus, Rncr3 KO mice also had smaller brains. Axons in the dentate gyrus of animals lacking Rnc3 also showed aberrant outgrowth. The tail suspension assay results in KO mice clasping their front and back limbs further demonstrated severe neuronal injury. Additionally, mouse models that mimic neurodegenerative diseases show this neuronal impairment. In conclusion, the KO of Rncr3 suggests that miR124 plays a significant role in murine brain development [158].

6. Dysregulation of miRNAs in embryonic development and disease

miRNAs have a role in differentially implantation and non-implantation of embryos; proliferative endometrium within women who face failed implantation and are used as a biomarker for embryo implantation [159], blastocyst development, and morula to blastocyst transition [160]; in short, they are involved in gametogenesis, embryogenesis, and mammalian fertility [12]. The role of short noncoding RNAs in gene expression is found in fully developed oocytes [160], ovaries, testis, and follicular fluid. During gametogenesis, miRNA are involved in the division and differentiation of stem cell and cell proliferation [84, 161] proved by knockout of Dicer1 gene [162] and AGO2 [163]. miRNA is involved in follicular development [164] through the regulation of vitamins, methionine, lipids, and cysteine [165]. miRNA has its role in testis development [166] as it is evident by its presence in Sertoli cells [167] and proved by the Dicer and Drosha development [166]. miRNA-1285, miRNA-762, miRNA-638, miRNA-15a,miR-29b,miR-34b and miR-34c has its role in spermatogenesis [168]. miRNA are also found in body fluids such as serum, saliva, urine, and blood plasma in the form of exosome or apoptotic bodies with the playing role as lipid or protein carriers [169, 170].

6.1 The effects of miRNA dysregulation on embryonic development and disease

miRNA can has deregulatory functions by destroying the translation process of protein making from mRNA [171]. This dysregulation has negative effects on embryonic development through having impact on stem cell division, cell proliferation, and on early or late embryonic development [161]. These negative effects lead to deleterious diseases such as tumors and cancers at the embryonic stage [172]. A total of 156 mature miRNAs are involved in embryonic cancerous diseases out of which colon cancer is greater in number [173]. Most importantly, miR-290–295, miR-302–367, miR-17–92, and miR-106b–25 are the set of miRNAs that are the controlling centers of embryonic diseases. Recent researches has also shown the effect of miRNA on protein output [174]. miRNA-383 is involved in infertility of men, maturation arrest, hyperactive proliferation of germ cells, G1-phase arrest, suppression of proliferation, and induction of apoptosis, and this effect is increased by the downregulation of IRF1. As described earlier, Dicer1, along with miRNA, has significant role in spermatogenesis. The Dicer1 deletion in the testis leads to defective spermatogenesis and during proliferation arrested at early differentiation. Only miRNA-372 and 373 are seen in defective spermatogenesis. This all would happen during early spermatogenesis stages, which can result in germ cell death [175]. miRNA expression is seen during different embryonic stages in mammals and thus can act as a biomarker for the selection of embryos for successful *in vitro* pregnancy. MiRNA is involved in polycystic ovary syndrome [176]. Recently shown that miRNA-25 is involved in cardiomyocytes oxidative stress and apoptosis in human ovarian cancer [177, 178]. miRNA-181 is involved early embryonic developmental stages and regulates nucleoplasmin 2 protein [179]. miRNA-196a encode transcriptional factors through HOX genes necessary for embryonic development [180]. Polymorphism miRNA-196aCC found in aborted fetuses [123]. miRNA-302 is found in blastocyst in high quantity. It is linked with cell programming [181]. miRNA-370 regulated DNA methyltransferase 3a expression [182].

6.2 miRNAs in developmental cancer

Different types of these small noncoding miRNAs are involved in different types of cancers [172], such as miR-15a is involved in leukemia [172], miR-145 in colorectal cancer [183], miR-21 in breast cancer [184], and let-7 of miRNA in lung cancer [172]. These dysregulated miRNAs can be oncogenes or tumor suppressors [172]. miRNA 17–92 are highly expressed in cancerous tissues in humans [174]. Amplification overexpression and depletion of miRNA lead to cancer [185]. miRNA are seen in genome effected with cancer [169]. miRNA is highly stable during extreme conditions of temperature and environment [186, 187]; so, miRNA is used as a biomarker for cancer such as lung, prostate, and colorectal cancer [184, 187]. miRNA expression gives 60% surety of cancer by analyzing serum, while 100% specificity is based on the effectiveness of miRNA [9]. Mutations, deletions, amplifications, epigenetic silencing, and dysregulation of transcriptional factors in miRNA loci are various processes that contribute to changes in the miRNA genome leading to cancer. The dysregulated miRNA genes are managed by protein-coding oncogenes or tumor suppressor gene. The tumor suppressor miRNA genes are B cell leukemia, lymphoma 2, MYC, RAS, phosphatase and tensin homolog (PTEN), p27, p57, and tissue inhibitor of metalloproteinases 3 (TIMP3). The silencing of miRNA genes gives epigenetic alterations that result in the silencing of tumor-suppressing genes. For example, the deletion of miRNA-29 causes DNA methyltransferase overexpression and results in tumor suppressor-silenced gene. The dysregulated miRNA can be used as biomarker for anticancer therapies [188].

Early B-cell lymphoma stages have greater levels of miRNA-21 expression; miRNA-200 and miRNA-17 are functionally active in cancer. The alternated expressions of miRNA in cancer were first described and studied by Chalin et al. [189]. miRNA-34a expression is reduced in breast, colon, and lung cancer [190]. The miRNA-331 expression is reduced in nasopharyngeal carcinoma. MiRNA-21, miRNA-155, miRNA-17b, and miRNA-106b has oncogenesis roles in hematological malignancies, B-cell lymphoma, breast, lung, and metastatic breast cancer, respectively [191]. Limited oxygen supply and hypoxia regulate the function of miRNA [192]. Altered miRNA genomic expressions in cancer arise from genomic variation. For instance, B-cell chronic lymphocytic leukemia has a high frequency of deletions at the chromosomal locus of the miR-15/miR-16 cluster (CLL) [193]. For instance, B-cell chronic lymphocytic leukemia frequently has genomic deletions at the miR-15/miR-16 cluster locus (CLL) [194].

Introduction of single nucleotide polymorphism (SNP) into miRNA sequence itself results in miRNA dysregulation both in epigenetic and at genetic level that leads to breast cancer [195]. A total of 50% of miRNA genes are involved in deletion, amplification, and translocations implicated tumorigenesis [196]. In fact, it is observed that 755 of miRNA genome is involved in breast cancer-affected copy number variations [196]. miRNA silencing *via* hypermethylation is observed in breast cancer with the discovery of miRNA-125b promotor in invasive breast cancer. It proves poor survival due to deletion of derepression of target gene ETSI. miRNA-335 gene is seen in breast cancer involving in suppressor of invasion, metastatic colonization, and tumor reinitiation [196]. The epigenetic discovery of miRNA discovered the new epigenetic-based cancer therapies. Mutations by XPO5 were seen during breast cancer, which is responsible for miRNA transportation to the cytoplasm. Both XOP5 and miRNA are responsible for breast cancer resulting in impaired RNA processing [197]. Genetic and epigenetic XOP5 is involved in breast cancer, while mutants are involved in the chances to get cancer [198]. While all this process, hypermethylation reduces the risk of cancer. Polymorphism, along with miRNA, is also responsible for breast cancer [199]. miRNAs are regulated by various signals released by progesterone and estrogen during breast cancer [200]. These signals dysregulate the miRNA.

7. Conclusion and prospects

MiRNAs control the physiological and developmental processes in a variety of animals, including embryos, by preventing the target genes' translation or degrading their activity. Endogenous tiny noncoding RNAs, which range in size from prokaryotes to eukaryotes, play a significant role in the transcriptional and post-transcriptional control of many genes. They control the complicated gene expression connected to the intricate process of embryogenesis throughout embryonic development. Further, study is, therefore, necessary to understand how miRNAs regulate gene expression at the posttranscriptional level, especially given that they are the most significant posttranscriptional regulators recently identified.

miRNA is playing its role in cell differentiation, proliferation, apoptosis, tumorigenesis, developmental timing, *in vitro* embryo development, and host-pathogen interaction. miRNA is itself control by various mechanism to maintain basic functions such as SNP, miRNA editing, methylation, and circadian clock.

Proven by 24,000 peer-reviewed articles, miRNA has its role in cancer. miRNA is involved in breast cancer through non-targeting hyaluronic acid. The dysregulated miRNA is involved in the malignant phenotype of cancer cells. MiRNA has its role as a therapeutic agent. The combination of small anticancer drugs and miRNA is effective in the treatment of cancer. Reducing drug resistance, increasing apoptosis and autophagy, reducing tumor angiogenesis, and preventing the production of drug efflux are all ways to reverse the EMT are some of the useful advantages of miRNA therapies over the traditional convention ones. The use of targeted miRNA is also proved effective in therapies.

MiRNA is important in gametogenesis and spermatogenesis. During the preimplantation stage of development, the embryo releases miRNA in extracellular vesicle. miRNA is involved in oogenesis, spermatogenesis, and early embryogenesis. This also includes several proteins of miRNA, including Dicer, Drosha, and DiGeorge. Dicer1 has generalized functions in handling other short RNAs, particularly siRNAs; therefore, it is not exclusive to miRNAs. miRNA is also involved in abortion and infertility. It is seen that the expose of varies to external environment can bring changes in oogenesis in relation of miRNA. miRNA-regulated gene is important in normal spermatogenesis. The activity of miRNA changes in the testis with spermatogenesis development.

Acknowledgements

Authors are grateful to the Vice Chancellors of the University of Narowal, the University of Punjab, and the University of Okara for providing support to this study.

Conflict of interest

The authors declare no conflict of interest.

Author details

Muhammad Babar Khawar[1†*], Huma Rasheed[1†], Ali Afzal[2†], Muddasir Hassan Abbasi[3*] and Nadeem Sheikh[4*]

1 Applied Molecular Biology and Biomedicine Lab, Department of Zoology, University of Narowal, Narowal, Pakistan

2 Molecular Medicine and Cancer Therapeutics Lab, Department of Zoology, Faculty of Science and Technology, University of Central Punjab, Lahore, Pakistan

3 Department of Zoology, University of Okara, Punjab, Pakistan

4 Cell and Molecular Biology Lab, Institute of Zoology, University of the Punjab, Lahore, Pakistan

*Address all correspondence to: babar.khawar@uon.edu.pk; dr.muddasir@uo.edu.pk and nadeem.zool@pu.edu.pk

† These authors contributed equally.

References

[1] Bartel DP. MicroRNAs: Genomics, biogenesis, mechanism, and function. Cell. 2004;**116**(2):281-297

[2] Prather R, First N. A review of early mouse embryogenesis and its applications to domestic species. Journal of Animal Science. 1988;**66**(10):2626-2635

[3] Yang Y, Bai W, Zhang L, Yin G, Wang X, Wang J, et al. Determination of microRNAs in mouse preimplantation embryos by microarray. Developmental Dynamics: An Official Publication of the American Association of Anatomists. 2008;**237**(9):2315-2327

[4] Suh N, Baehner L, Moltzahn F, Melton C, Shenoy A, Chen J, et al. MicroRNA function is globally suppressed in mouse oocytes and early embryos. Current Biology. 2010;**20**(3):271-277

[5] Yang Q, Lin J, Liu M, Li R, Tian B, Zhang X, et al. Highly sensitive sequencing reveals dynamic modifications and activities of small RNAs in mouse oocytes and early embryos. Science Advances. 2016;**2**(6):e1501482

[6] Duranthon V, Watson AJ, Lonergan P. Preimplantation embryo programming: Transcription, epigenetics, and culture environment. Reproduction (Cambridge, England). 2008;**135**(2):141

[7] Bernstein E, Kim SY, Carmell MA, Murchison EP, Alcorn H, Li MZ, et al. Dicer is essential for mouse development. Nature Genetics. 2003;**35**(3):215-217

[8] Alisch RS, Jin P, Epstein M, Caspary T, Warren ST. Argonaute2 is essential for mammalian gastrulation and proper mesoderm formation. PLoS Genetics. 2007;**3**(12):e227

[9] Tang F, Kaneda M, O'Carroll D, Hajkova P, Barton SC, Sun YA, et al. Maternal microRNAs are essential for mouse zygotic development. Genes & Development. 2007;**21**(6):644-648

[10] Cheong AW, Pang RT, Liu W-M, Kottawatta KSA, Lee K-F, Yeung WS. MicroRNA let-7a and dicer are important in the activation and implantation of delayed implanting mouse embryos. Human Reproduction. 2014;**29**(4):750-762

[11] Houbaviy HB, Murray MF, Sharp PA. Embryonic stem cell-specific MicroRNAs. Developmental Cell. 2003;**5**(2):351-358

[12] Salilew-Wondim D, Gebremedhn S, Hoelker M, Tholen E, Hailay T, Tesfaye D. The role of micrornas in mammalian fertility: From gametogenesis to embryo implantation. International Journal of Molecular Sciences. 2020;**21**(2):585

[13] Spruce T, Pernaute B, Di-Gregorio A, Cobb BS, Merkenschlager M, Manzanares M, et al. An early developmental role for miRNAs in the maintenance of extraembryonic stem cells in the mouse embryo. Developmental Cell. 2010;**19**(2):207-219

[14] Morales-Prieto D, Chaiwangyen W, Ospina-Prieto S, Schneider U, Herrmann J, Gruhn B, et al. MicroRNA expression profiles of trophoblastic cells. Placenta. 2012;**33**(9):725-734

[15] Foshay KM, Gallicano GI. miR-17 family miRNAs are expressed during early mammalian development and regulate stem cell differentiation.

Developmental Biology. 2009;**326**(2): 431-443

[16] Kim J, Lee J, Jun JH. Identification of differentially expressed microRNAs in outgrowth embryos compared with blastocysts and non-outgrowth embryos in mice. Reproduction, Fertility and Development. 2019;**31**(4):645-657

[17] Hinnebusch AG, Ivanov IP, Sonenberg N. Translational control by 5′-untranslated regions of eukaryotic mRNAs. Science. 2016;**352**(6292):1413-1416

[18] Date R. Bradyrhizobium effectiveness responses in Stylosanthes hamata and S. seabrana. Tropical Grasslands. 2010;**44**(3):141-157

[19] Ambros V. microRNAs: Tiny regulators with great potential. Cell. 2001;**107**(7):823-826

[20] Treiber T, Treiber N, Meister G. Regulation of microRNA biogenesis and its crosstalk with other cellular pathways. Nature Reviews Molecular cell biology. 2019;**20**(1):5-20

[21] Filipowicz W, Bhattacharyya SN, Sonenberg N. Mechanisms of post-transcriptional regulation by microRNAs: Are the answers in sight? Nature Reviews Genetics. 2008;**9**(2):102-114

[22] Kura B, Szeiffova Bacova B, Kalocayova B, Sykora M, Slezak J. Oxidative stress-responsive microRNAs in heart injury. International Journal of Molecular Sciences. 2020;**21**(1):358

[23] Stavast CJ, Erkeland SJ. The non-canonical aspects of microRNAs: Many roads to gene regulation. Cell. 2019;**8**(11):1465

[24] Salim U, Kumar A, Kulshreshtha R, Vivekanandan P. Biogenesis, characterization, and functions of mirtrons. Wiley Interdisciplinary Reviews: RNA. 2022;**13**(1):e1680

[25] Olena A, Patton J. MiRNA biogenesis and function. In: Singh S, Rameshwar P, editors. MicroRNA in Development and in the Progression of Cancer. New York, NY: Springer; 2014. DOI: 10.1007/978-1-4899-8065-6_1

[26] Lee Y, Kim M, Han J, Yeom KH, Lee S, Baek SH, et al. MicroRNA genes are transcribed by RNA polymerase II. The EMBO Journal. 2004;**23**(20):4051-4060

[27] Michlewski G, Cáceres JF. Post-transcriptional control of miRNA biogenesis. RNA. 2019;**25**(1):1-16

[28] Liu J, Zhou F, Guan Y, Meng F, Zhao Z, Su Q, et al. The biogenesis of miRNAs and their role in the development of amyotrophic lateral sclerosis. Cell. 2022;**11**(3):572

[29] Ha M, Kim VN. Regulation of microRNA biogenesis. Nature reviews Molecular Cell Biology. 2014;**15**(8):509-524

[30] Nguyen DX, Bos PD, Massagué J. Metastasis: From dissemination to organ-specific colonization. Nature Reviews Cancer. 2009;**9**(4):274-284

[31] Siomi H, Siomi MC. Posttranscriptional regulation of microRNA biogenesis in animals. Molecular Cell. 2010;**38**(3):323-332

[32] Kataoka N, Fujita M, Ohno M. Functional association of the microprocessor complex with the spliceosome. Molecular and Cellular Biology. 2009;**29**(12):3243-3254

[33] Cullen BR. Transcription and processing of human microRNA

precursors. Molecular Cell. 2004;**16**(6):861-865

[34] Annese T, Tamma R, De Giorgis M, Ribatti D. microRNAs biogenesis, functions and role in tumor angiogenesis. Frontiers in Oncology. 2020;**10**:581007

[35] Noland CL, Doudna JA. Multiple sensors ensure guide strand selection in human RNAi pathways. RNA. 2013;**19**(5):639-648

[36] Ender C, Krek A, Friedländer MR, Beitzinger M, Weinmann L, Chen W, et al. A human snoRNA with microRNA-like functions. Molecular Cell. 2008;**32**(4):519-528

[37] Natarelli L, Weber C. A non-canonical link between non-coding RNAs and cardiovascular diseases. Biomedicine. 2022;**10**(2):445

[38] Cole C, Sobala A, Lu C, Thatcher SR, Bowman A, Brown JW, et al. Filtering of deep sequencing data reveals the existence of abundant dicer-dependent small RNAs derived from tRNAs. RNA. 2009;**15**(12):2147-2160

[39] Liu B, Cao J, Wang X, Guo C, Liu Y, Wang T. Deciphering the tRNA-derived small RNAs: Origin, development, and future. Cell Death & Disease. 2021;**13**(1):24

[40] Maute RL, Schneider C, Sumazin P, Holmes A, Califano A, Basso K, et al. tRNA-derived microRNA modulates proliferation and the DNA damage response and is down-regulated in B cell lymphoma. Proceedings of the National Academy of Sciences. 2013;**110**(4):1404-1409

[41] Grishok A, Pasquinelli AE, Conte D, Li N, Parrish S, Ha I, et al. Genes and mechanisms related to RNA interference regulate expression of the small temporal RNAs that control C. elegans developmental timing. Cell. 2001;**106**(1):23-34

[42] Fukunaga R, Han BW, Hung J-H, Xu J, Weng Z, Zamore PD. Dicer partner proteins tune the length of mature miRNAs in flies and mammals. Cell. 2012;**151**(3):533-546

[43] Jonas S, Izaurralde E. Towards a molecular understanding of microRNA-mediated gene silencing. Nature reviews Genetics. 2015;**16**(7):421-433

[44] Lee FC, Ule J. Advances in CLIP technologies for studies of protein-RNA interactions. Molecular Cell. 2018;**69**(3):354-369

[45] Gebert LF, MacRae IJ. Regulation of microRNA function in animals. Nature Reviews Molecular Cell Biology. 2019;**20**(1):21-37

[46] Golden RJ, Chen B, Li T, Braun J, Manjunath H, Chen X, et al. An Argonaute phosphorylation cycle promotes microRNA-mediated silencing. Nature. 2017;**542**(7640):197-202

[47] Quévillon Huberdeau M, Zeitler DM, Hauptmann J, Bruckmann A, Fressigné L, Danner J, et al. Phosphorylation of Argonaute proteins affects mRNA binding and is essential for micro RNA-guided gene silencing in vivo. The EMBO Journal. 2017;**36**(14):2088-2106

[48] Bridge KS, Shah KM, Li Y, Foxler DE, Wong SC, Miller DC, et al. Argonaute utilization for miRNA silencing is determined by phosphorylation-dependent recruitment of LIM-domain-containing proteins. Cell Reports. 2017;**20**(1):173-187

[49] Freimer JW, Krishnakumar R, Cook MS, Blelloch R. Expression of

alternative Ago2 isoform associated with loss of microRNA-driven translational repression in mouse oocytes. Current Biology. 2018;**28**(2):296-302.e3

[50] Kobayashi H, Shoji K, Kiyokawa K, Negishi L, Tomari Y. Iruka eliminates dysfunctional argonaute by selective ubiquitination of its empty state. Molecular Cell. 2019;**73**(1):119-29.e5

[51] Bell CE, Calder MD, Watson AJ. Genomic RNA profiling and the programme controlling preimplantation mammalian development. Molecular Human Reproduction. 2008;**14**(12):691-701

[52] Gerri C, Menchero S, Mahadevaiah SK, Turner JM, Niakan KK. Human embryogenesis: A comparative perspective. Annual Review of Cell and Developmental Biology. 2020;**36**:411-440

[53] Kaczmarek MM, Najmula J, Guzewska MM, Przygrodzka E. miRNAs in the peri-implantation period: Contribution to embryo–maternal communication in pigs. International Journal of Molecular Sciences. 2020;**21**(6):2229

[54] Liang J, Wang S, Wang Z. Role of microRNAs in embryo implantation. Reproductive Biology and Endocrinology. 2017;**15**:1-11

[55] Enders AC, Schlafke S, Hendrickx AG. Differentiation of the embryonic disc, amnion, and yolk sac in the rhesus monkey. American Journal of Anatomy. 1986;**177**(2):161-185

[56] Bentin-Ley U, Sjögren A, Nilsson L, Hamberger L, Larsen J, Horn T. Presence of uterine pinopodes at the embryo–endometrial interface during human implantation in vitro. Human Reproduction. 1999;**14**(2):515-520

[57] Chobotova K, Spyropoulou I, Carver J, Manek S, Heath JK, Gullick WJ, et al. Heparin-binding epidermal growth factor and its receptor ErbB4 mediate implantation of the human blastocyst. Mechanisms of Development. 2002;**119**(2):137-144

[58] Hamatani T, Carter MG, Sharov AA, Ko MS. Dynamics of global gene expression changes during mouse preimplantation development. Developmental Cell. 2004;**6**(1):117-131

[59] Song H, Lim H, Das SK, Paria BC, Dey SK. Dysregulation of EGF family of growth factors and COX-2 in the uterus during the preattachment and attachment reactions of the blastocyst with the luminal epithelium correlates with implantation failure in LIF-deficient mice. Molecular Endocrinology. 2000;**14**(8):1147-1161

[60] Mohamed OA, Jonnaert M, Labelle-Dumais C, Kuroda K, Clarke HJ, Dufort D. Uterine Wnt/β-catenin signaling is required for implantation. Proceedings of the National Academy of Sciences. 2005;**102**(24):8579-8584

[61] Gross N, Kropp J, Khatib H. MicroRNA signaling in embryo development. Biology. 2017;**6**(3):34

[62] Tian S, Su X, Qi L, Jin X-H, Hu Y, Wang C-L, et al. MiR-143 and rat embryo implantation. Biochimica et Biophysica Acta (BBA)-General Subjects. 2015;**1850**(4):708-721

[63] Li R, He J, Chen X, Ding Y, Wang Y, Long C, et al. Mmu-miR-193 is involved in embryo implantation in mouse uterus by regulating GRB7 gene expression. Reproductive Sciences. 2014;**21**:733-742

[64] Chu B, Zhong L, Dou S, Wang J, Li J, Wang M, et al. miRNA-181 regulates embryo implantation in mice through targeting leukemia inhibitory factor. Journal of Molecular Cell Biology. 2015;7(1):12-22

[65] Liang S, Nie Z-W, Guo J, Niu Y-J, Shin K-T, Ock SA, et al. Overexpression of microRNA-29b decreases expression of DNA methyltransferases and improves quality of the blastocysts derived from somatic cell nuclear transfer in cattle. Microscopy and Microanalysis. 2018;**24**(1):29-37

[66] Li Z, Jia J, Gou J, Tong A, Liu X, Zhao X, et al. Mmu-miR-126a-3p plays a role in murine embryo implantation by regulating Itga11. Reproductive Biomedicine Online. 2015;**31**(3):384-393

[67] Shen L-J, He J-L, Yang D-H, Ding Y-B, Chen X-M, Geng Y-Q, et al. Mmu-microRNA-200a overexpression leads to implantation defect by targeting phosphatase and tensin homolog in mouse uterus. Reproductive Sciences. 2013;**20**(12):1518-1528

[68] Bloor D, Metcalfe A, Rutherford A, Brison D, Kimber S. Expression of cell adhesion molecules during human preimplantation embryo development. MHR: Basic Science of Reproductive Medicine. 2002;**8**(3):237-245

[69] Ichimura A, Ruike Y, Terasawa K, Tsujimoto G. miRNAs and regulation of cell signaling. The FEBS Journal. 2011;**278**(10):1610-1618

[70] Inui M, Martello G, Piccolo S. MicroRNA control of signal transduction. Nature Reviews Molecular Cell Biology. 2010;**11**(4):252-263

[71] Terasawa K, Ichimura A, Sato F, Shimizu K, Tsujimoto G. Sustained activation of ERK1/2 by NGF induces microRNA-221 and 222 in PC12 cells. The FEBS Journal. 2009;**276**(12):3269-3276

[72] Taganov KD, Boldin MP, Chang K-J, Baltimore D. NF-κB-dependent induction of microRNA miR-146, an inhibitor targeted to signaling proteins of innate immune responses. Proceedings of the National Academy of Sciences. 2006;**103**(33):12481-12486

[73] Martello G, Zacchigna L, Inui M, Montagner M, Adorno M, Mamidi A, et al. MicroRNA control of nodal signalling. Nature. 2007;**449**(7159):183-188

[74] Rossant J, Tam PPL. Blastocyst lineage formation, early embryonic asymmetries and axis patterning in the mouse. Development. 2009;**136**(5): 701-713. DOI: 10.1242/dev.017178

[75] Okae H, Toh H, Sato T, Hiura H, Takahashi S, Shirane K, et al. Derivation of human trophoblast stem cells. Cell Stem Cell. 2018;**22**(1):50-63.e6

[76] Turnpenny L, Brickwood S, Spalluto CM, Piper K, Cameron IT, Wilson DI, et al. Derivation of human embryonic germ cells: An alternative source of pluripotent stem cells. Stem Cells. 2003;**21**(5):598-609

[77] Blakeley P, Fogarty NM, Del Valle I, Wamaitha SE, Hu TX, Elder K, et al. Defining the three cell lineages of the human blastocyst by single-cell RNA-seq. Development. 2015;**142**(18):3151-3165

[78] Georgantas RW III, Hildreth R, Morisot S, Alder J. Liu C-g, Heimfeld S, et al. CD34+ hematopoietic stem-progenitor cell microRNA expression and function: A circuit diagram of differentiation control. Proceedings of the National Academy of Sciences. 2007;**104**(8):2750-2755

[79] Gangaraju VK, Lin H. MicroRNAs: Key regulators of stem cells. Nature Reviews Molecular Cell Biology. 2009;**10**(2):116-125

[80] Barroso-delJesus A, Romero-López C, Lucena-Aguilar G, Melen GJ, Sanchez L, Ligero G, et al. Embryonic

stem cell-specific miR302-367 cluster: Human gene structure and functional characterization of its core promoter. Molecular and Cellular Biology. 2008;**28**(21):6609-6619

[81] Yeung F, Chung E, Guess MG, Bell ML, Leinwand LA. Myh7b/miR-499 gene expression is transcriptionally regulated by MRFs and Eos. Nucleic Acids Research. 2012;**40**(15):7303-7318

[82] Chen J-F, Tao Y, Li J, Deng Z, Yan Z, Xiao X, et al. microRNA-1 and microRNA-206 regulate skeletal muscle satellite cell proliferation and differentiation by repressing Pax7. Journal of Cell Biology. 2010;**190**(5):867-879

[83] Vecchione A, Croce CM. Apoptomirs: Small molecules have gained the license to kill. Endocrine-Related Cancer. 2010;**17**(1):F37

[84] Brennecke J, Hipfner DR, Stark A, Russell RB, Cohen SM. Bantam encodes a developmentally regulated microRNA that controls cell proliferation and regulates the proapoptotic gene hid in drosophila. Cell. 2003;**113**(1):25-36

[85] Zhang X, Zhang X, Hu S, Zheng M, Zhang J, Zhao J, et al. Identification of miRNA-7 by genome-wide analysis as a critical sensitizer for TRAIL-induced apoptosis in glioblastoma cells. Nucleic Acids Research. 2017;**45**(10):5930-5944

[86] Favreau AJ, Shaffiey F, Cross E, Sathyanarayana P. Mir-590 is a novel STAT5 regulated oncogenic miRNA and targets FasL in acute myeloid leukemia. Blood. 2013;**122**(21):3811

[87] Wang P, Zhuang L, Zhang J, Fan J, Luo J, Chen H, et al. The serum miR-21 level serves as a predictor for the chemosensitivity of advanced pancreatic cancer, and miR-21 expression confers chemoresistance by targeting FasL. Molecular Oncology. 2013;**7**(3):334-345

[88] Huang G, Nishimoto K, Zhou Z, Hughes D, Kleinerman ES. miR-20a encoded by the miR-17-92 cluster increases the metastatic potential of osteosarcoma cells by regulating Fas ExpressionmiR-20a regulates Fas and osteosarcoma metastatic potential. Cancer Research. 2012;**72**(4):908-916

[89] Huang X, Xiao S, Zhu X, Yu Y, Cao M, Zhang X, et al. miR-196b-5p-mediated downregulation of FAS promotes NSCLC progression by activating IL6-STAT3 signaling. Cell Death & Disease. 2020;**11**(9):785

[90] Yamada N, Noguchi S, Kumazaki M, Shinohara H, Miki K, Naoe T, et al. Epigenetic regulation of microRNA-128a expression contributes to the apoptosis-resistance of human T-cell leukaemia jurkat cells by modulating expression of fas-associated protein with death domain (FADD). Biochimica et Biophysica Acta (BBA)-molecular. Cell Research. 2014;**1843**(3):590-602

[91] Zhang J, Du Y, Wu C, Ren X, Ti X, Shi J, et al. Curcumin promotes apoptosis in human lung adenocarcinoma cells through miR-186* signaling pathway. Oncology Reports. 2010;**24**(5):1217-1223

[92] Czochor JR, Glazer PM. microRNAs in cancer cell response to ionizing radiation. Antioxidants & Redox Signaling. 2014;**21**(2):293-312

[93] Yin W, Chen J, Wang G, Zhang D. MicroRNA-106b functions as an oncogene and regulates tumor viability and metastasis by targeting LARP4B in prostate cancer. Molecular Medicine Reports. 2019;**20**(2):951-958

[94] Park JK, Doseff AI, Schmittgen TD. MicroRNAs targeting caspase-3 and-7 in

PANC-1 cells. International Journal of Molecular Sciences. 2018;**19**(4):1206

[95] Wang Y, Zhang S, Bao H, Mu S, Zhang B, Ma H, et al. MicroRNA-365 promotes lung carcinogenesis by downregulating the USP33/SLIT2/ROBO1 signalling pathway. Cancer Cell International. 2018;**18**(1):1-14

[96] Li Q, Ren P, Shi P, Chen Y, Xiang F, Zhang L, et al. MicroRNA-148a promotes apoptosis and suppresses growth of breast cancer cells by targeting B-cell lymphoma 2. Anti-Cancer Drugs. 2017;**28**(6):588-595

[97] Zhuo C, Jiang R, Lin X, Shao M. LncRNA H19 inhibits autophagy by epigenetically silencing of DIRAS3 in diabetic cardiomyopathy. Oncotarget. 2017;**8**(1):1429

[98] Zhang Y, Schiff D, Park D, Abounader R. MicroRNA-608 and microRNA-34a regulate chordoma malignancy by targeting EGFR, Bcl-xL and MET. PLoS One. 2014;**9**(3):e91546

[99] Chen Q, Xu J, Li L, Li H, Mao S, Zhang F, et al. MicroRNA-23a/b and microRNA-27a/b suppress Apaf-1 protein and alleviate hypoxia-induced neuronal apoptosis. Cell Death & Disease. 2014;**5**(3):e1132-e

[100] Ma L, Zheng Y, Tang X, Gao H, Liu N, Gao Y, et al. miR-21-3p inhibits autophagy of bovine granulosa cells by targeting VEGFA via PI3K/AKT signaling. Reproduction. 2019;**158**(5):441-452

[101] Li Z, Li W. miR-383 inhibits proliferation of granulosa cells by down-regulation of cell cycle-related proteins in mice. Xi bao yu fen zi Mian yi xue za zhi= Chinese journal of. Cellular and Molecular Immunology. 2019;**35**(6):518-525

[102] Gebremedhn S, Salilew-Wondim D, Hoelker M, Rings F, Neuhoff C, Tholen E, et al. MicroRNA-183-96-182 cluster regulates bovine granulosa cell proliferation and cell cycle transition by coordinately targeting FOXO1. Biology of Reproduction. 2016;**94(6):127**:1-11

[103] Liu J, Li X, Yao Y, Li Q, Pan Z, Li Q. miR-1275 controls granulosa cell apoptosis and estradiol synthesis by impairing LRH-1/CYP19A1 axis. Biochimica et Biophysica Acta (BBA)-Gene Regulatory Mechanisms. 2018;**1861**(3):246-257

[104] Liu J, Yao W, Yao Y, Du X, Zhou J, Ma B, et al. MiR-92a inhibits porcine ovarian granulosa cell apoptosis by targeting Smad7 gene. FEBS Letters. 2014;**588**(23):4497-4503

[105] Sirotkin V, A, Kisova G, Brenaut P, Ovcharenko D, Grossmann R, Mlyncek M. Involvement of microRNA Mir15a in control of human ovarian granulosa cell proliferation, apoptosis, steroidogenesis, and response to FSH. Microrna. 2014;**3**(1):29-36

[106] Mohammed BT, Sontakke SD, Ioannidis J, Duncan WC, Donadeu FX. The adequate corpus luteum: miR-96 promotes luteal cell survival and progesterone production. The Journal of Clinical Endocrinology & Metabolism. 2017;**102**(7):2188-2198

[107] Yuan X, Deng X, Zhou X, Zhang A, Xing Y, Zhang Z, et al. MiR-126-3p promotes the cell proliferation and inhibits the cell apoptosis by targeting TSC1 in the porcine granulosa cells. In Vitro Cellular & Developmental Biology-Animal. 2018;**54**:715-724

[108] Zhang M, Zhang Q, Hu Y, Xu L, Jiang Y, Zhang C, et al. miR-181a increases FoxO1 acetylation and promotes granulosa cell apoptosis via

SIRT1 downregulation. Cell Death & Disease. 2017;**8**(10):e3088-e

[109] Nie M, Yu S, Peng S, Fang Y, Wang H, Yang X. miR-23a and miR-27a promote human granulosa cell apoptosis by targeting SMAD5. Biology of Reproduction. 2015;**93**(4):98, 1-10

[110] Geng Y, Sui C, Xun Y, Lai Q, Jin L. MiRNA-99a can regulate proliferation and apoptosis of human granulosa cells via targeting IGF-1R in polycystic ovary syndrome. Journal of Assisted Reproduction and Genetics. 2019;**36**:211-221

[111] Yao L, Li M, Hu J, Wang W, Gao M. MiRNA-335-5p negatively regulates granulosa cell proliferation via SGK3 in PCOS. Reproduction. 2018;**156**(5):439-449

[112] Yao Y, Niu J, Sizhu S, Li B, Chen Y, Li R, et al. microRNA-125b regulates apoptosis by targeting bone morphogenetic protein receptor 1B in yak granulosa cells. DNA and Cell Biology. 2018;**37**(11):878-887

[113] Zhang L, Zhang X, Zhang X, Lu Y, Li L, Cui S. MiRNA-143 mediates the proliferative signaling pathway of FSH and regulates estradiol production. Journal of Endocrinology. 2023;**234**(1):1-14. DOI: 10.1530/JOE-16-0488

[114] Giraldez AJ, Cinalli RM, Glasner ME, Enright AJ, Thomson JM, Baskerville S, et al. MicroRNAs regulate brain morphogenesis in zebrafish. Science. 2005;**308**(5723):833-838

[115] Kanellopoulou C, Muljo SA, Kung AL, Ganesan S, Drapkin R, Jenuwein T, et al. Dicer-deficient mouse embryonic stem cells are defective in differentiation and centromeric silencing. Genes & Development. 2005;**19**(4):489-501

[116] Cobb BS, Nesterova TB, Thompson E, Hertweck A, O'Connor E, Godwin J, et al. T cell lineage choice and differentiation in the absence of the RNase III enzyme dicer. The Journal of Experimental Medicine. 2005;**201**(9):1367-1373

[117] Muljo SA, Ansel KM, Kanellopoulou C, Livingston DM, Rao A, Rajewsky K. Aberrant T cell differentiation in the absence of dicer. The Journal of Experimental Medicine. 2005;**202**(2):261-269

[118] Chen C-Z, Li L, Lodish HF, Bartel DP. MicroRNAs modulate hematopoietic lineage differentiation. Science. 2004;**303**(5654):83-86

[119] Harfe BD, McManus MT, Mansfield JH, Hornstein E, Tabin CJ. The RNaseIII enzyme dicer is required for morphogenesis but not patterning of the vertebrate limb. Proceedings of the National Academy of Sciences. 2005;**102**(31):10898-10903

[120] Lee RC, Ambros V. An extensive class of small RNAs in Caenorhabditis elegans. Science. 2001;**294**(5543):862-864

[121] Zhao Y, Samal E, Srivastava D. Serum response factor regulates a muscle-specific microRNA that targets Hand2 during cardiogenesis. Nature. 2005;**436**(7048):214-220

[122] Harvey RP. Nk-2homeobox genes and heart development. Developmental Biology. 1996;**178**(2):203-216

[123] Ahn DH, Rah H, Choi YK, Jeon YJ, Min KT, Kwack K, et al. Association of the miR-146aC> G, miR-149T> C, miR-196a2T> C, and miR-499A> G polymorphisms with gastric cancer risk and survival in the Korean

population. Molecular Carcinogenesis. 2013;**52**(S1):39-51

[124] Ciruna B, Rossant J. FGF signaling regulates mesoderm cell fate specification and morphogenetic movement at the primitive streak. Developmental Cell. 2001;**1**(1):37-49

[125] Harvey RP. Patterning the vertebrate heart. Nature Reviews Genetics. 2002;**3**(7):544-556

[126] Andrée B, Duprez D, Vorbusch B, Arnold H-H, Brand T. BMP-2 induces ectopic expression of cardiac lineage markers and interferes with somite formation in chicken embryos. Mechanisms of Development. 1998;**70**(1-2):119-131

[127] Kitajima S, Takagi A, Inoue T, Saga Y. MesP1 and MesP2 are essential for the development of cardiac mesoderm. Development. 2000;**127**(15):3215-3226

[128] Bondue A, Lapouge G, Paulissen C, Semeraro C, Iacovino M, Kyba M, et al. Mesp1 acts as a master regulator of multipotent cardiovascular progenitor specification. Cell Stem Cell. 2008;**3**(1):69-84

[129] Lough J, Barron M, Brogley M, Sugi Y, Bolender DL, Zhu X. Combined BMP-2 and FGF-4, but neither factor alone, induces cardiogenesis in non-precardiac embryonic mesoderm. Developmental Biology. 1996;**178**(1):198-202

[130] Brade T, Männer J, Kühl M. The role of Wnt signalling in cardiac development and tissue remodelling in the mature heart. Cardiovascular Research. 2006;**72**(2):198-209

[131] Bodmer R, Jan LY, Jan Y. A new homeobox-containing gene, msh-2, is transiently expressed early during mesoderm formation of drosophila. Development. 1990;**110**(3):661-669

[132] Kasahara H, Bartunkova S, Schinke M, Tanaka M, Izumo S. Cardiac and extracardiac expression of Csx/Nkx2. 5 homeodomain protein. Circulation Research. 1998;**82**(9):936-946

[133] Lyons I, Parsons LM, Hartley L, Li R, Andrews JE, Robb L, et al. Myogenic and morphogenetic defects in the heart tubes of murine embryos lacking the homeo box gene Nkx2-5. Genes & Development. 1995;**9**(13):1654-1666

[134] Tanaka M, Chen Z, Bartunkova S, Yamasaki N, Izumo S. The cardiac homeobox gene Csx/Nkx2. 5 lies genetically upstream of multiple genes essential for heart development. Development. 1999;**126**(6):1269-1280

[135] Newman CS, Krieg PA. Tinman-related genes expressed during heart development in Xenopus. Developmental Genetics. 1998;**22**(3):230-238

[136] Cohen ED, Wang Z, Lepore JJ, Lu MM, Taketo MM, Epstein DJ, et al. Wnt/β-catenin signaling promotes expansion of Isl-1–positive cardiac progenitor cells through regulation of FGF signaling. The Journal of Clinical Investigation. 2007;**117**(7):1794-1804

[137] Lin L, Cui L, Zhou W, Dufort D, Zhang X, Cai C-L, et al. β-Catenin directly regulates Islet1 expression in cardiovascular progenitors and is required for multiple aspects of cardiogenesis. Proceedings of the National Academy of Sciences. 2007;**104**(22):9313-9318

[138] Tzahor E. Wnt/β-catenin signaling and cardiogenesis: Timing does matter. Developmental Cell. 2007;**13**(1):10-13

[139] Challice C, Virágh S. The architectural development of the early mammalian heart. Tissue and Cell. 1974;**6**(3):447-462

[140] Van Mierop L, Gessner IH. The morphologic development of the sinoatrial node in the mouse. The American Journal of Cardiology. 1970;**25**(2):204-212

[141] Buckingham M, Meilhac S, Zaffran S. Building the mammalian heart from two sources of myocardial cells. Nature Reviews Genetics. 2005;**6**(11):826-835

[142] Warburton D, El-Hashash A, Carraro G, Tiozzo C, Sala F, Rogers O, et al. Lung organogenesis. Current Topics in Developmental Biology. 2010;**90**:73-158

[143] Jiang Z, Cushing L, Ai X, Lü J. miR-326 is downstream of sonic hedgehog signaling and regulates the expression of Gli2 and smoothened. American Journal of Respiratory Cell and Molecular Biology. 2014;**51**(2):273-283

[144] Carraro G, Shrestha A, Rostkovius J, Contreras A, Chao C-M, El Agha E, et al. miR-142-3p balances proliferation and differentiation of mesenchymal cells during lung development. Development. 2014;**141**(6):1272-1281

[145] Carraro G, El-Hashash A, Guidolin D, Tiozzo C, Turcatel G, Young BM, et al. miR-17 family of microRNAs controls FGF10-mediated embryonic lung epithelial branching morphogenesis through MAPK14 and STAT3 regulation of E-cadherin distribution. Developmental Biology. 2009;**333**(2):238-250

[146] Sanford EL, Choy KW, Donahoe PK, Tracy AA, Hila R, Loscertales M, et al. MiR-449a affects epithelial proliferation during the pseudoglandular and canalicular phases of avian and mammal lung development. PLoS One. 2016;**11**(2):e0149425

[147] Sun YF, Kan Q, Yang Y, Zhang YH, Shen JX, Zhang C, et al. Knockout of microRNA-26a promotes lung development and pulmonary surfactant synthesis. Molecular Medicine Reports. 2018;**17**(4):5988-5995

[148] Wang Y, Frank DB, Morley MP, Zhou S, Wang X, Lu MM, et al. HDAC3-dependent epigenetic pathway controls lung alveolar epithelial cell remodeling and spreading via miR-17-92 and TGF-β signaling regulation. Developmental Cell. 2016;**36**(3):303-315

[149] Bhaskaran M, Wang Y, Zhang H, Weng T, Baviskar P, Guo Y, et al. MicroRNA-127 modulates fetal lung development. Physiological Genomics. 2009;**37**(3):268-278

[150] Lal CV, Olave N, Travers C, Rezonzew G, Dolma K, Simpson A, et al. Exosomal microRNA predicts and protects against severe bronchopulmonary dysplasia in extremely premature infants. JCI Insight. 2018;**3**(5)

[151] Zhang Y, Coarfa C, Dong X, Jiang W, Hayward-Piatkovskyi B, Gleghorn JP, et al. MicroRNA-30a as a candidate underlying sex-specific differences in neonatal hyperoxic lung injury: Implications for BPD. American Journal of Physiology-Lung Cellular and Molecular Physiology. 2019;**316**(1):L144-LL56

[152] Åkerblom M, Sachdeva R, Quintino L, Wettergren EE, Chapman KZ, Manfre G, et al. Visualization and genetic modification of resident brain microglia using lentiviral vectors regulated by microRNA-9. Nature Communications. 2013;**4**(1):1770

[153] Cheng L-C, Pastrana E, Tavazoie M, Doetsch F. miR-124 regulates adult

neurogenesis in the subventricular zone stem cell niche. Nature Neuroscience. 2009;**12**(4):399-408

[154] Åkerblom M, Sachdeva R, Barde I, Verp S, Gentner B, Trono D, et al. MicroRNA-124 is a subventricular zone neuronal fate determinant. Journal of Neuroscience. 2012;**32**(26):8879-8889

[155] Krichevsky AM, Sonntag K-C, Isacson O, Kosik KS. Specific microRNAs modulate embryonic stem cell–derived neurogenesis. Stem Cells. 2006;**24**(4):857-864

[156] Lund E, Guttinger S, Calado A, Dahlberg JE, Kutay U. Nuclear export of microRNA precursors. Science. 2004;**303**(5654):95-98

[157] Xia H, Cheung WK, Ng SS, Jiang X, Jiang S, Sze J, et al. Loss of brain-enriched miR-124 microRNA enhances stem-like traits and invasiveness of glioma cells. Journal of Biological Chemistry. 2012;**287**(13):9962-9971

[158] Sanuki R, Onishi A, Koike C, Muramatsu R, Watanabe S, Muranishi Y, et al. miR-124a is required for hippocampal axogenesis and retinal cone survival through Lhx2 suppression. Nature Neuroscience. 2011;**14**(9):1125-1134

[159] Paul AB, Sadek ST, Mahesan AM. The role of microRNAs in human embryo implantation: A review. Journal of Assisted Reproduction and Genetics. 2019;**36**:179-187

[160] Paulson EE, Fishman EL, Ma J, Schultz RM, Ross PJ. Embryonic microRNAs are essential for bovine preimplantation embryo development. Proceedings of the National Academy of Sciences. 2022;**119**(45):e2212942119

[161] Hatfield S, Shcherbata H, Fischer K, Nakahara K, Carthew R, Ruohola-Baker H. Stem cell division is regulated by the microRNA pathway. Nature. 2005;**435**(7044):974-978

[162] Lei L, Jin S, Gonzalez G, Behringer RR, Woodruff TK. The regulatory role of dicer in folliculogenesis in mice. Molecular and Cellular Endocrinology. 2010;**315**(1-2):63-73

[163] Kaneda M, Tang F, O'Carroll D, Lao K, Surani MA. Essential role for Argonaute2 protein in mouse oogenesis. Epigenetics & Chromatin. 2009;**2**:1-8

[164] Gebremedhn S, Salilew-Wondim D, Ahmad I, Sahadevan S, Hossain MM, Hoelker M, et al. MicroRNA expression profile in bovine granulosa cells of preovulatory dominant and subordinate follicles during the late follicular phase of the estrous cycle. PLoS One. 2015;**10**(5):e0125912

[165] Salilew-Wondim D, Ahmad I, Gebremedhn S, Sahadevan S, Hossain MM, Rings F, et al. The expression pattern of microRNAs in granulosa cells of subordinate and dominant follicles during the early luteal phase of the bovine estrous cycle. PLoS One. 2014;**9**(9):e106795

[166] Yuan S, Schuster A, Tang C, Yu T, Ortogero N, Bao J, et al. Sperm-borne miRNAs and endo-siRNAs are important for fertilization and preimplantation embryonic development. Development. 2016;**143**(4):635-647

[167] Wainwright EN, Jorgensen JS, Kim Y, Truong V, Bagheri-Fam S, Davidson T, et al. SOX9 regulates microRNA miR-202-5p/3p expression during mouse testis differentiation. Biology of Reproduction. 2013;**89**(2):34, 1-12

[168] Gao H, Wen H, Cao C, Dong D, Yang C, Xie S, et al. Overexpression of microRNA-10a in germ cells causes male

infertility by targeting Rad51 in mouse and human. Frontiers in Physiology. 2019;**10**:765

[169] Wang Z, Yao H, Lin S, Zhu X, Shen Z, Lu G, et al. Transcriptional and epigenetic regulation of human microRNAs. Cancer Letters. 2013;**331**(1):1-10

[170] Rayner KJ, Hennessy EJ. Thematic review series: Functional regulation of lipid homeostasis by microRNA: Extracellular communication via microRNA: Lipid particles have a new message. Journal of Lipid Research. 2013;**54**(5):1174

[171] Narry KV, Nam J-W. Genomics of microRNA. TRENDS in Genetics. 2006;**22**(3):165-173

[172] Eder M, Scherr M. MicroRNA and lung cancer. New England Journal of Medicine. 2005;**352**(23):2446-2448

[173] Monzo M, Navarro A, Bandres E, Artells R, Moreno I, Gel B, et al. Overlapping expression of microRNAs in human embryonic colon and colorectal cancer. Cell Research. 2008;**18**(8):823-833

[174] Vidigal JA, Ventura A. Embryonic stem cell miRNAs and their roles in development and disease. In: Seminars in Cancer Biology. Vol. 22. Issue 5-6. Academic Press, Elsevier; 2012. pp. 428-436

[175] Lian J, Tian H, Liu L, Zhang X, Li W, Deng Y, et al. Downregulation of microRNA-383 is associated with male infertility and promotes testicular embryonal carcinoma cell proliferation by targeting IRF1. Cell Death & Disease. 2010;**1**(11):e94-e

[176] Grive KJ. Pathways coordinating oocyte attrition and abundance during mammalian ovarian reserve establishment. Molecular Reproduction and Development. 2020;**87**(8):843-856

[177] Varga ZV, Kupai K, Szűcs G, Gáspár R, Pálóczi J, Faragó N, et al. MicroRNA-25-dependent up-regulation of NADPH oxidase 4 (NOX4) mediates hypercholesterolemia-induced oxidative/nitrative stress and subsequent dysfunction in the heart. Journal of Molecular and Cellular Cardiology. 2013;**62**:111-121

[178] Zhang H, Zuo Z, Lu X, Wang L, Wang H, Zhu Z. MiR-25 regulates apoptosis by targeting Bim in human ovarian cancer. Oncology Reports. 2012;**27**(2):594-598

[179] Lingenfelter BM, Tripurani SK, Tejomurtula J, Smith GW, Yao J. Molecular cloning and expression of bovine nucleoplasmin 2 (NPM2): A maternal effect gene regulated by miR-181a. Reproductive Biology and Endocrinology. 2011;**9**:1-9

[180] Chen C, Zhang Y, Zhang L, Weakley SM, Yao Q. MicroRNA-196: Critical roles and clinical applications in development and cancer. Journal of Cellular and Molecular Medicine. 2011;**15**(1):14-23

[181] Lee MR, Prasain N, Chae H-D, Kim Y-J, Mantel C, Yoder MC, et al. Epigenetic regulation of NANOG by miR-302 cluster-MBD2 completes induced pluripotent stem cell reprogramming. Stem Cells. 2013;**31**(4):666-681

[182] Qi L, Hongjuan H, Ning G, Zhengbin H, Yanjiang X, Tiebo Z, et al. miR-370 is stage-specifically expressed during mouse embryonic development and regulates Dnmt3a. FEBS Letters. 2013;**587**(6):775-781

[183] Schetter AJ, Leung SY, Sohn JJ, Zanetti KA, Bowman ED, Yanaihara N,

et al. MicroRNA expression profiles associated with prognosis and therapeutic outcome in colon adenocarcinoma. Journal of the American Medical Association. 2008;**299**(4):425-436

[184] Chan JA, Krichevsky AM, Kosik KS. MicroRNA-21 is an antiapoptotic factor in human glioblastoma cells. Cancer Research. 2005;**65**(14):6029-6033

[185] Lin S, Gregory RI. MicroRNA biogenesis pathways in cancer. Nature Reviews Cancer. 2015;**15**(6):321-333

[186] Chen K, Boettiger A, Moffitt J, Wang S, Zhuang X. Spatially resolved, highly multiplexed RNA profiling in single cells. Science. 2015;**348**(6233)

[187] Stupar RM. Into the wild: The soybean genome meets its undomesticated relative. Proceedings of the National Academy of Sciences. 2010;**107**(51):21947-21948

[188] Croce CM. Causes and consequences of microRNA dysregulation in cancer. Nature Reviews Genetics. 2009;**10**(10):704-714

[189] Calin GA, Dumitru CD, Shimizu M, Bichi R, Zupo S, Noch E, et al. Frequent deletions and down-regulation of micro-RNA genes miR15 and miR16 at 13q14 in chronic lymphocytic leukemia. Proceedings of the National Academy of Sciences. 2002;**99**(24):15524-15529

[190] Xuefang Z, Ruinian Z, Liji J, Chun Z, Qiaolan Z, Jun J, et al. RETRACTED: miR-331-3p inhibits proliferation and promotes apoptosis of nasopharyngeal carcinoma cells by targeting elf4B-PI3K-AKT pathway. Technology in Cancer Research & Treatment. 2020;**19**:1533033819892251

[191] Medina PP, Nolde M, Slack FJ. OncomiR addiction in an in vivo model of microRNA-21-induced pre-B-cell lymphoma. Nature. 2010;**467**(7311):86-90

[192] Franovic A, Gunaratnam L, Smith K, Robert I, Patten D, Lee S. Translational up-regulation of the EGFR by tumor hypoxia provides a nonmutational explanation for its overexpression in human cancer. Proceedings of the National Academy of Sciences. 2007;**104**(32):13092-13097

[193] Allegra D, Bilan V, Garding A, Döhner H, Stilgenbauer S, Kuchenbauer F, et al. Defective DROSHA processing contributes to downregulation of MiR-15/−16 in chronic lymphocytic leukemia. Leukemia. 2014;**28**(1):98-107

[194] Zhao JL, Starczynowski DT. Role of microRNA-146a in normal and malignant hematopoietic stem cell function. Frontiers in Genetics. 2014;**5**:219

[195] Melo SA, Esteller M. Dysregulation of microRNAs in cancer: Playing with fire. FEBS Letters. 2011;**585**(13):2087-2099

[196] Zhang L, Huang J, Yang N, Greshock J, Megraw MS, Giannakakis A, et al. microRNAs exhibit high frequency genomic alterations in human cancer. Proceedings of the National Academy of Sciences. 2006;**103**(24):9136-9141

[197] Melo SA, Moutinho C, Ropero S, Calin GA, Rossi S, Spizzo R, et al. A genetic defect in exportin-5 traps precursor microRNAs in the nucleus of cancer cells. Cancer Cell. 2010;**18**(4):303-315

[198] Leaderer D, Hoffman AE, Zheng T, Fu A, Weidhaas J, Paranjape T, et al. Genetic and epigenetic association studies suggest a role of microRNA biogenesis gene exportin-5 (XPO5) in breast tumorigenesis. International

Journal of Molecular Epidemiology and Genetics. 2011;**2**(1):9

[199] Sung H, Jeon S, Lee K-M, Han S, Song M, Choi J-Y, et al. Common genetic polymorphisms of microRNA biogenesis pathway genes and breast cancer survival. BMC Cancer. 2012;**12**(1):1-12

[200] Yang Z, Wang L. Regulation of microRNA expression and function by nuclear receptor signaling. Cell & Bioscience. 2011;**1**(1):1-9

Chapter 7

Photosynthesis in Plants Undergoing Silencing

Christos Kotakis

Abstract

RNA silencing shares common features among different eukaryotes. However little is known about the metabolic consequences of this mechanism relate to the (plant) cell homeostasis. Here, we probe the chlroroplast bioenergetics in transgenic plants undergoing silencing. An increased capacity for non-photochemical energy quenching followed by a limiting photosystem II functionality characterize the photosynthesis of silenced cells compared to non-silenced ones. These alterations are accompanied by a significant up-regulation of photosystem I, providing evidence for active cyclic electron flow in silencing conditions. The biological significance of our results is discussed related to possible energetic inter-communication between photosynthesis and RNA silencing.

Keywords: ATP, cyclic electron flow, photosynthesis, photosystem I, RNA silencing mechanism

1. Introduction

RNA-mediated gene silencing pathways are responsible for the regulation of plant development, buffering of the genome integrity and for the effective response of plant organism to various stressful a/−biotic factors (reviewed by [1]). Currently, it has been identified a positive effect of light intensity on the induction and spread of spontaneous RNA silencing in transgenic *N. benthamiana* plants, indicating possible implication of photosynthesis in this phenomenon [2]. The objective of this study is to investigate the impact of the silencing signal on the plant's energetic repertoire. For that reason a detailed photosynthetic characterization was performed in transgenic tissues that are characterized by a dynamic transition from non-silenced to silenced state. We tried to characterize possible mechanistic aspects of the interplay between RNA silencing and light intensity in plants [3], via biophysical, biochemical and molecular data. This piece of work could promote the understanding of the crosstalk between RNA silencing mechanism and photosynthesis in transgenic plants. Therefore, we believe that our experimental findings would supply a creative motivation for further investigation in the area of other RNA-mediated gene silencing pathways, concerning the plant physiology.

2. Methods

2.1 Plant material

N. benthamiana transgenic lines that were used in this study are described in Kotakis et al. [2]. More specifically, GFP transgenic lines (line 6.4 and 5.1.1) that initiate local as well as systemic silencing spontaneously and the NIB-transgenic line 20-1A1 that undergoes also silencing, were used. Plants were grown as described in Kotakis et al. [2], under a white (PAR: 400–700 nm) light intensity of 130 μmol m^{-2} s^{-1}. Systemically fully silenced leaves from non-silenced ones; they are termed 'silenced' or non-silenced' respectively inside the text, were discriminated phenotypically as described in [2]. All data are derived from experiments in line 6.4, otherwise the specific plant line used is cited inside the text. The constitutively GFP-expressing line 16c was used for agroinfiltration of a hairpin *GFP* construct in order to induce systemic silencing as described in Kościańska et al. [4].

2.2 *In vivo* chlorophyll fluorescence measurements

The parameters studied in the light curves were quantified as described in Lichtenthaler et al. [5] using a PAM-210 fluorometer (Waltz, Germany). The fluorescence data concerning maximum photosynthetic performance [6] were obtained using a Handy-PEA fluorometer (Hansatech, UK) by consequent OJIP-analysis with Biolyser 4.0 software. Estimation of the relative plastoquinone (PQ) pool reduction as well as DCMU treatment were performed according to Toth et al. [7]. The capacity for state transitions and cyclic electron flow in examining leaf types was calculated according to Bellafiore et al. [8] and Munne-Bosch et al. [9], respectively.

2.3 Thylakoids isolation and polarography

Isolation of functional thylakoids from leaves and linear electron flow rate (ETR) estimation were performed according to Casazza et al. [10]. Activity of photosystem (PS) I & PSII was estimated as described in Romanowska et al. [11].

2.4 Determination of adenylates and reducing equivalents

ATP and NADPH production from thylakoid membranes were estimated as described in Casazza et al. [10]. ATP and ADP content in leaves were extracted and quantified by HPLC as described in Andronis and Roubelakis-Angelakis [12].

2.5 Putrescine quantification

Putrescine extraction from leaves and following quantification were performed in accordance to Navakoudis et al. [13].

2.6 Immunoblotting

Phosphorylation of LHCII was determined by western blotting as described in Bellafiore et al. [8].

2.7 Quantitative real-time PCR analysis

The mRNA levels of chloroplastic *PSaA* (GI:81238323) and *PSbA* (GI:76559634) genes that encode reactions centers of PSI and PSII respectively, were quantified by qPCR analysis. RNA extraction and qPCR was performed as previously described in Kotakis et al. [2] and the sequence of primer pairs used, is given in **Table A1**.

2.8 Method of sampling, repetition of experiments and statistics

For the *in vivo* measurements 5 leaves of similar growth, angle and distance of their surface to the light source, from corresponding different plants were used in each category studied. For all the biochemical and molecular approaches, the experiments were repeated at least 3 times in a pooled material from 5 different plants. All the plots represent average ± standard deviations. Significance of differences in the various parameters plotted with bars, was assessed by One-way ANOVA ($P < 0.05$), (SPSS 17.0, SPSS Inc., USA). Single asterisks denote statistically significant differences between wt tissue and either one of non-silenced or silenced tissue. The double asterisks denote statistically significant differences between non-silenced tissue and silenced tissue. In the **Figures 2E** and **3B**, the presence of asterisks denotes statistical significance of the difference displaying, compared to the zero-change level and in the case of **Figure A3B**, they stand for statistical significance between the two treatments.

The etymology of some abbreviations used in the text, is provided in **Table A2**.

3. Results and discussion

3.1 Photosynthesis is differentiated in tissues undergoing silencing

In this study, firstly we attempted to correlate basic bioenergetic parameters that characterize the photosynthetic process, with the RNA silencing conditions. Towards this aim we used chlorophyll a fluorescence induction kinetics as a probe for quantifying PSII behavior in transgenic *N. bethanmiana* plants that are functionally related to the presence of the RNA silencing signal. Non-silenced transgenic tissues as well as wild-type (wt) samples were served as controls. At least two different transgenic lines (three for the silenced- and two for the non-silenced state) were used in this study in order to eliminate possible masking in our results from factors such as accidental insertional effects in genomic loci related to the plant physiology and specificities due to the transgene nature. The data obtained by this analysis denoted that similar bioenergetic profiles (**Figures 1A** and **A1A**) characterize the transgenic lines studied with an active RNA silencing mechanism and though quite different in comparison to the corresponding controls. Also, it has to be emphasized that two different transgenic lines related to non-silenced state shared a high degree of similarity in such type of measurements (**Figure A1B**). According to these data sets, it seems that the photosynthesis is remarkably differentiated in comparison to either non-silenced or wt conditions. More specifically, a limiting potential for reduction of PSII acceptor side is reported in silencing conditions as it can be confirmed by a diminishing gradient of the maximum chlorophyll a fluorescence plateau from the non-silenced to silenced state (**Figure 1B**).

Next, we decided to continue our study, determining the steady-state photosynthesis, since the preliminary experiments had revealed similar bioenergetic principles, independently the transgenic line studied. For this purpose, we chose one

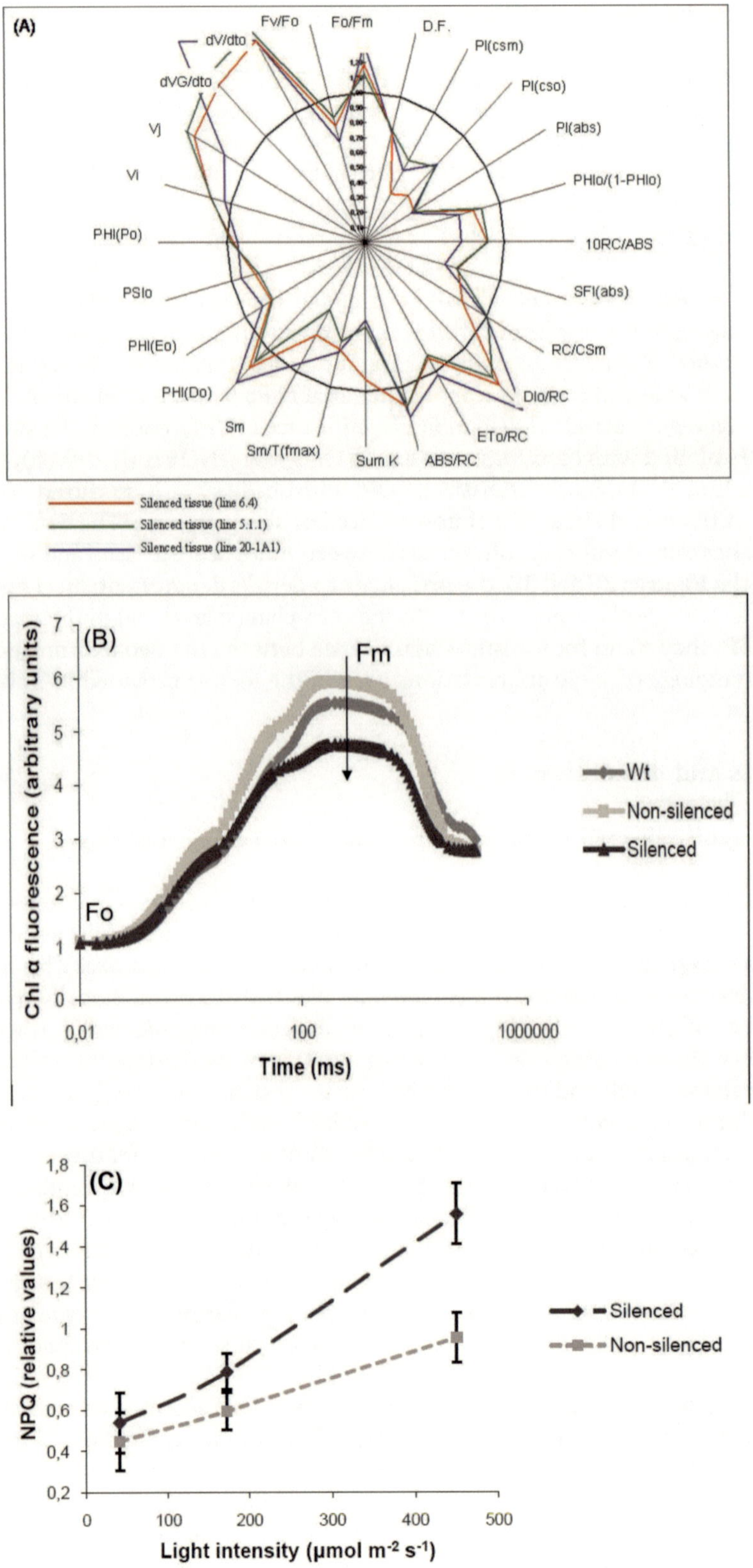
(A)
Fv/Fo
Fo/Fm
D.F.
PI(csm)
PI(cso)
PI(abs)
PHIo/(1-PHIo)
10RC/ABS
SFI(abs)
RC/CSm
DIo/RC
ETo/RC
ABS/RC
Sum K
Sm/T(fmax)
Sm
PHI(Do)
PHI(Eo)
PSIo
PHI(Po)
Vi
Vj
dVG/dto
dV/dto
Silenced tissue (line 6.4)
Silenced tissue (line 5.1.1)
Silenced tissue (line 20-1A1)
(B)
Fm
Fo
Chl α fluorescence (arbitrary units)
Time (ms)
0,01
100
1000000
Wt
Non-silenced
Silenced
(C)
NPQ (relative values)
Light intensity (μmol m-2 s-1)
Silenced
Non-silenced

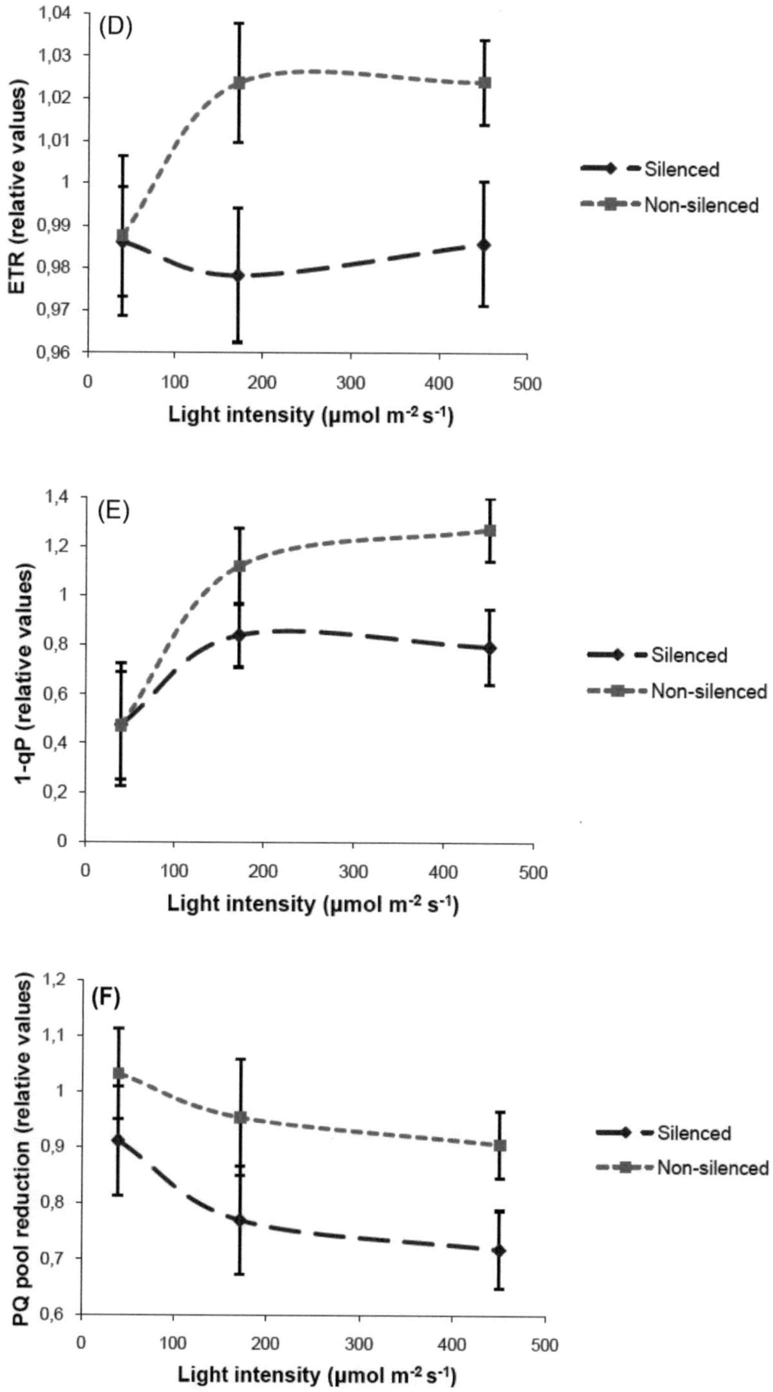

Figure 1.
The photosynthetic function of cells where silencing is established, is altered in comparison to non-silenced cells. A: Bioenergetic profiles are displaying basic parameters of maximum photosynthetic acclimation in the examining tissues. The control condition (non-silenced tissue) is presented as circle. B: Representative chlorophyll fluorescence induction curves, normalized by the Fo. The time in the x-axis is presented in a logarithmic scale. The arrow depicts the Fm plateau. Light curves for C: NPQ, D: ETR, E: 1-qP and for F: PQ pool reduction, in different light intensities. Transgenic leaf values normalized over corresponding wt values in the parameters of light curves. NPQ; non-photochemical quenching, ETR: Electron transport rate, 1-qP; excitation pressure of PSII.

transgenic line that has advantage to display silenced and non-silenced plant areas at the same time, with a high frequency in the initiation and spread of silencing events [2]. As we can conclude from the respective light curves, silenced tissues are characterized by a deficient PSII energy perception as indicated by up to 67% higher capacity for NPQ formation than the non-silenced ones (**Figure 1C**). Furthermore, the coherent utilization of light energy absorbed by PSII is considerably limiting, regarding the linear electron flow, (**Figure 1D**) in silencing conditions. Next, the corresponding PSII excitation pressure levels remain quite low (**Figure 1E**), most probably due to the inefficient generation of reduced PQ pool that was found to follow concomitant trace (**Figure 1F**), accordingly to the low electron flow rates. Taken together the above data, we can point out that the redox potential of chloroplasts that drives the linear electron flow to provide both ATP and NADPH is not efficient in silenced cells.

3.2 PSI plays a significant role in chloroplast bioenergetics of silenced cells

Drawing the outline up to now, the chemiosmotic capacity of the photosynthetic apparatus in silenced cells is significantly limiting because the availability of intermediate electron carriers at PSII acceptor side was found decreased compared to the non-silenced cells. The last is shown also by the statistically lower ETR, even normalized in a chlorophyll basis (**Figure A2**). By contrast it is arisen that possible adequate sinks of electrons may be managed as dissipating valves in the electron transport chain, when the redox poise of plastoquinone was reduced by inhibiting the electron flow at the Q_B binding site. Particularly, not only the reduced PQ pool remained low but also the 'silenced' values were maintained at significant high levels compared to the almost close to zero, 'non-silenced' ones (**Figure 2C**). So, we make the hypothesis of the probable existence of a strong source of electrons donation and it is proposed the switch to a different mode of electron flow (e.g. PSI-alternative route of electrons via PQ recycling) [14].

Then, we decided to resolve the composition and function of the thylakoid membrane in depth. Mechanisms such as state transitions and photosystem stoichiometry adjustment were needed to be elucidated by biochemical and molecular means, since both responses are functionally coupling with the redox state of the plastoquinone pool (reviewed by [15]) that found modified in silenced cells. The silenced tissues show a decreased, though not statistically significant, capacity for the migration of LHCII from state 1 to state 2 (**Figure 2A**), in comparison to the non-silenced ones. The tendency for light utilization preferably not by PSII in non-silenced cells, is accompanied by a strong phosphorylation of LHCII antenna (**Figure 2B**). These findings show an active state transition mechanism in cells that are prepared to perceive the silencing signal. On the other hand, the complete lack of corresponding signal in silencing conditions (**Figure 2B**), maybe underline that this type of regulation is not sufficient and/or inactivated since it is related to short-term time scales of response (reviewed by [16]) and LHCII kinase is activated by a specific range of PQ redox state [17], respectively. In this case, a second level of long-term regulation such as the photosystem stoichiometry adjustment is possibly needed. Actually, this organization strategy is reinforced as it is reflected by an approximate 1.5 fold increment to the PSI/PSII activity (**Figure 2D**) in the thylakoid membranes of tissues undergoing silencing in comparison to non-silenced ones. This functional PSI-enrichment in the chloroplasts of plant cells that are characterized by an metabolically active RNA silencing mechanism is entailed by a transcriptional up-regulation of the *PSaA* gene over 5-fold in the silenced tissues, with the mRNAs of *PSbA*, to be remained almost zero-changed (**Figure 2E**). The modified stoichiometry of photosystems even in the transcriptional level, comes to substantiate the rest body of evidence for a significant PSI involvement in the chloroplast bioenergetics, when the RNA silencing signal has been established.

3.3 Evidence for ATP requirement via cyclic electron flow in chloroplasts of cells, where silencing signal is established

The decreased PQ pool in reduced form, can be justified by the transcriptional PS stoichiometry [18] found when RNA silencing mechanism is activated. A switch to cyclic electron flow is quite possible that is compensating for the inefficient operation of the linear electron flow of chloroplasts there-in, since the first mode of flow is driven by the PSI alone, and it provides a mechanism whereby ATP production can be increased

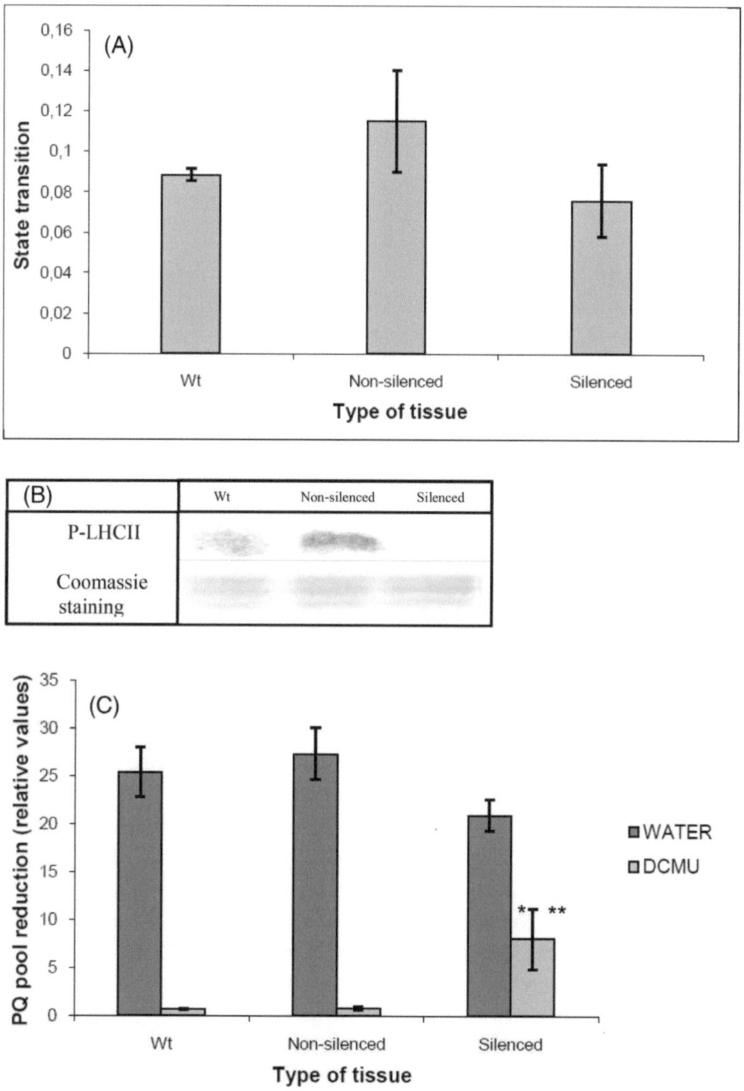

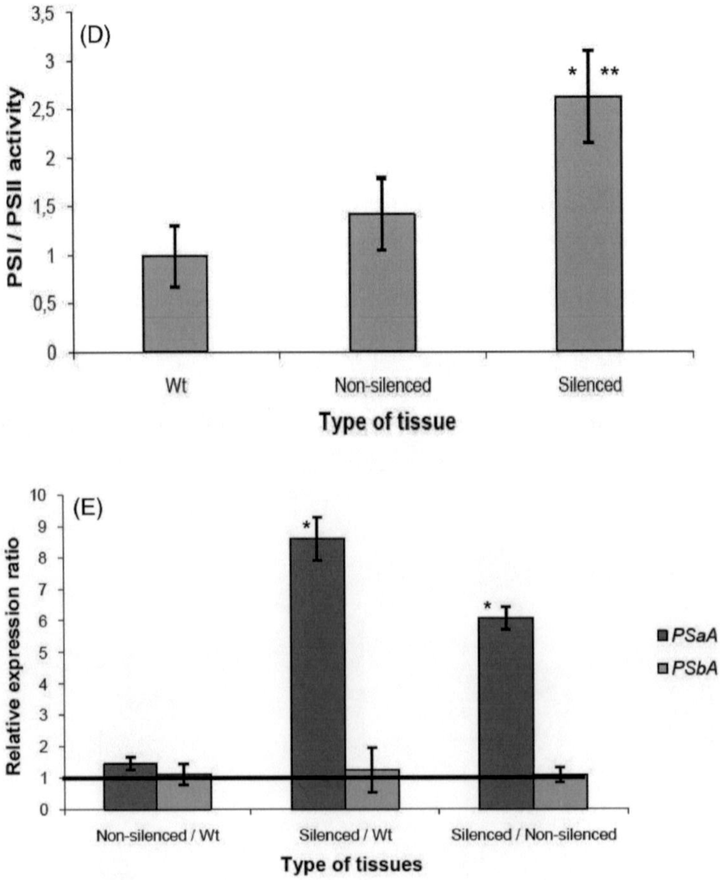

Figure 2.
Evidence for a significant PSI involvement in plant cells with the signature of RNA silencing. A: Capacity for transition from state 1 to state 2 and B: Phosphorylation levels of LHCII among the photosynthetic tissues studied. C: Quantification of the steady-state PQ pool in control (water) and oxidized (DCMU) conditions for the PSII acceptor side. D: Relative photosystem's activity (PSI/PSII) in thylakoid membranes of the corresponding leaf tissues. E: Relative expression ratio of PS genes, between the different leaf types examined. The horizontal line designates the zero-change level, (ratio one).

relative to NADPH (reviewed by [19]). Quantifying the capacity for cyclic electron flow, a statistical higher difference is exhibited in silenced tissues compared to the non-silenced ones (**Figure 3A**). This regulatory adaptation confirms that the metabolic conditions inside the silenced cells impose the necessity for proper adjustment of the ATP/NADPH ratio that is reported to follow more than 2-fold increment (**Figure 3B**). We further note the slightly increased ATP/NADPH ratio of non-silenced tissues in comparison to wt ones (**Figure 3B**) that may reflect the energetic requirements of the plant cell metabolism in order to sustain the transgene expression in a constitutive manner. The pattern of the altered chloroplastic contribution is clearly manifested in the ATP/ADP ratio of silenced cells (**Figure 3C**), validating the augmentation of ATP there-in. From all the above, it emerges that the mode of electron flow in the chloroplast of cells undergoing RNA silencing is different compared to cells that restore an energetic budget towards a continuous transgene expression. As a consequence, we could speculate that cyclic electron flow around PSI is accelerated in this type of tissues, providing additional ATP molecules in cells where RNA silencing is maintained.

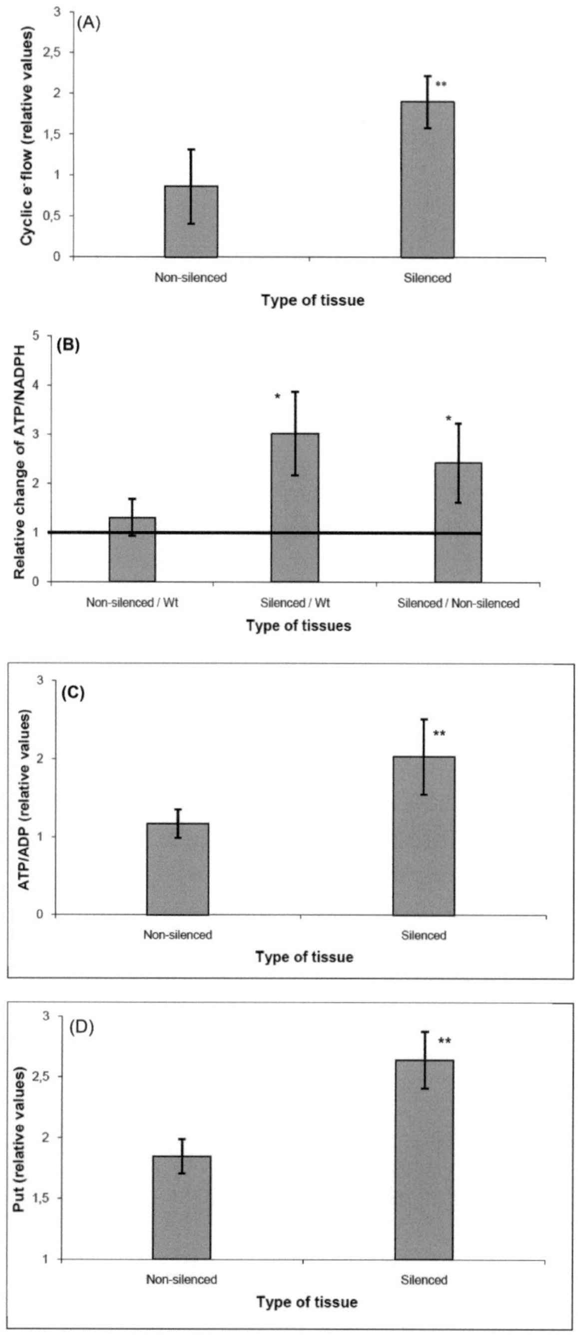

Figure 3.
The adenylate status is redistributed in tissues undergoing silencing, via an active cyclic electron flow. A: Capacity for cyclic electron flow and B: Relative change of chloroplastic ATP/NADPH ratio, between the different plant types, examined. The horizontal line designates the zero-change level. C: ATP/ADP ratio and D: Put content, in fully silenced leaf tissues and non-silenced ones. Data normalized over corresponding wt values in A, C and D.

Hence, an interplay between silencing and photosynthesis is addressed, taking also into account the differential response of RNA silencing to variations of light intensity [2]. To testify this notification, we tried to simulate HL-conditions by using the polyamine Put. The role of this biogenic amine is known currently, concerning the regulation of the photo-adaptive status of photosynthetic apparatus [13] via an increased chloroplastic ATP pool [20]. Based on the last, the content of Put was quantified in the examining tissues. Fully silenced leaves were found to contain higher levels of Put, compared to the non-silenced ones (**Figure 3D**). This result is in agreement with the rearrangement of the energy charge in plant cells' in silenced state. After that, non-silenced leaf tissues were treated by Put, creating an ATP-fuelled environment inside the corresponding cells. Daily monitoring of the silencing initiation and spread events was followed in the plant individuals studied. The scores obtained by Put-treated samples were higher (**Figure A3A**) and followed by earlier initiated phenotypes (**Figure A3B**) compared to the short-range silencing and systemic signal appearance, respectively in control conditions. The data derived by these time kinetics suggest that silencing performance could efficiently be regulated by bioenergetic manipulation, strengthening the possibility for inter-communication between silencing and photosynthesis.

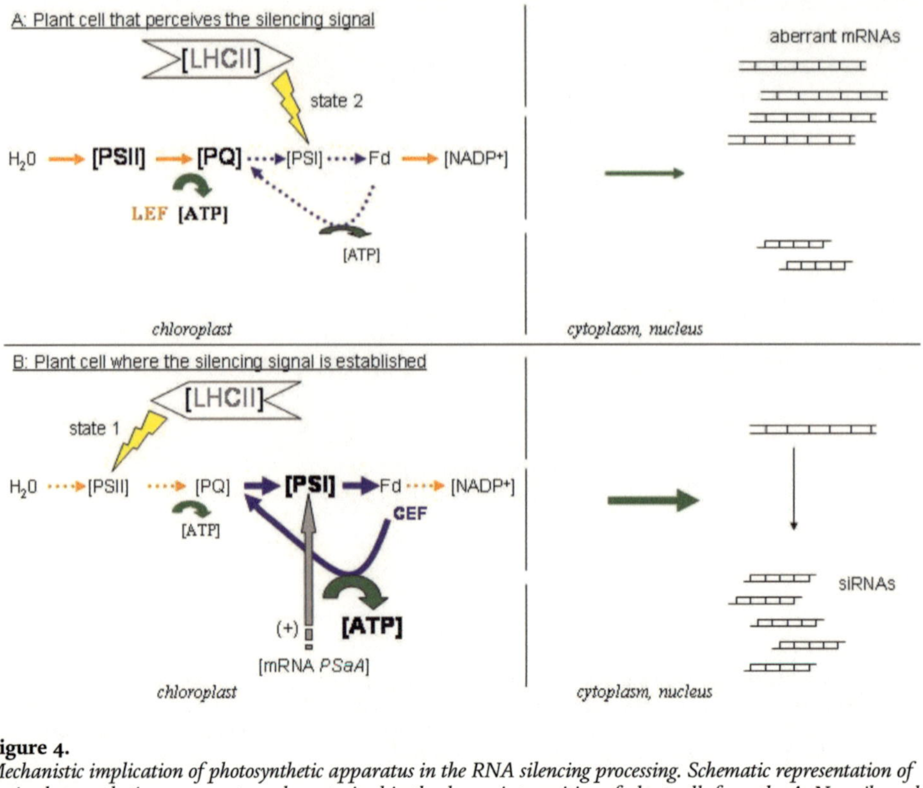

Figure 4.
Mechanistic implication of photosynthetic apparatus in the RNA silencing processing. Schematic representation of basic photosynthetic components as characterized in the dynamic transition of plant cells from the A: Non-silenced to B: Silenced state. Orange and blue arrows stand for the linear (LEF) and cyclic electron flow (CEF), respectively. The green ones stand for the ATP generation. The relative size of letters and arrows that depicted in the diagram, as well as the graphical (dis)- continuity of some arrows is indicative of their functional role inside the cell.

By conclusion, the structure and function of the photosynthetic machinery are differentiated in tissues undergoing silencing compared to tissues where silencing signal has not been established yet. The photosynthetic mechanism under silencing conditions is described by an enhancement of the non-photochemical quenching process of the energy captured by the light-harvesting antenna and a subsequent increased PSI/PSII activity. The PQ redox state act as a signal for the LHCII transition in state 2, in non-silenced cells (**Figure 4A**) and for a transcriptional up-regulation of *PSaA* gene in silenced ones. As a result, the onset of the cyclic electron flow is engaging and additional ATP molecules are likely provided in cell metabolism where silencing conditions are supported (**Figure 4B**). Taking into consideration our results, we envision that the ATP requirement of the biochemical pathway of RNA silencing mechanism, may be responsible for the changes observed. The ATP-dependent processing of dsRNA into siRNAs [21, 22] and the enhanced RISC activity in the presence of ATP [21] are some mechanistic aspects for an effective silencing mechanism. There is also evidence for a role of ATP in unwinding of siRNA duplex as well as in maintaining 5′ phosphate groups in a functional siRNA molecule [23].

It is claimed that plants grown under HL conditions, are characterized by higher availability of ATP pools in comparison to low light grown plants (reviewed by [24]). Hence, we are tempting to speculate that one part of these energetic equivalents is related with the increased frequency of silencing events as well as with an effective and conductive silencing signal, concerning the overall RNA silencing perception found in HL conditions [2]. Thus, cyclic electron flow is considered to constitute a protective mechanism against redox imbalances (e.g. stress conditions), acting synergistically with the antioxidant machinery for restoring the energetic and redox status of the plant cell (reviewed by [25]). So, it cannot be excluded that such micro-environmental conditions could probably be linked with the dynamic transition from a strong transgene expression to siRNAs 'burst' and it remains to be investigated.

3.4 Core finding and perspective

We report evidence that chloroplast bioenergetics is altered remarkably in silencing conditions and activation of transgene silencing imposes the photosystem I involvement in chloroplast energetics. The data propose bioenergetic manipulation as potential means of regulating silencing efficiency.

A. Appendix

Primer ID	Primer sequence 5′-3′
PSaA-FOR	GCTCTAGATGGCAGGGCTACTAGGA
PSaA-REV	CGAATTCAATGGTGATGGGCAATA
PSbA-FOR	GCTCTAGATTGACGGCAACTTCTGT
PSbA-REV	CGAATTCCCAAGGTCGCATACCCA

Table A1.
List of primer sequences used in quantitative real-time PCR assays.

Abbreviation	Etymology
DCMU	3-(3′,4′-dichlorophenyl) 1, 1′-dimethyl urea
Fo	Initial fluorescence value
Fm	Maximum fluorescence value
Fd	Ferredoxin
GFP	Green Fluorescent Protein
HL	High Light
HPLC	High-Performance Liquid Chromatography
LHCII	Light-Harvesting Complex II
NIB	*Plum Pox Virus* polymerase
PAR	Photosynthetically Active Radiation
PAM	Pulse Amplitude Fluorometry
Put	Putrescine
Q_B	Quinone B
RISC	RNA-Induced-Silencing-Complex
siRNA	Short-interfering RNA

Table A2.
List with some abbreviations that are cited inside the text and their corresponding etymology.

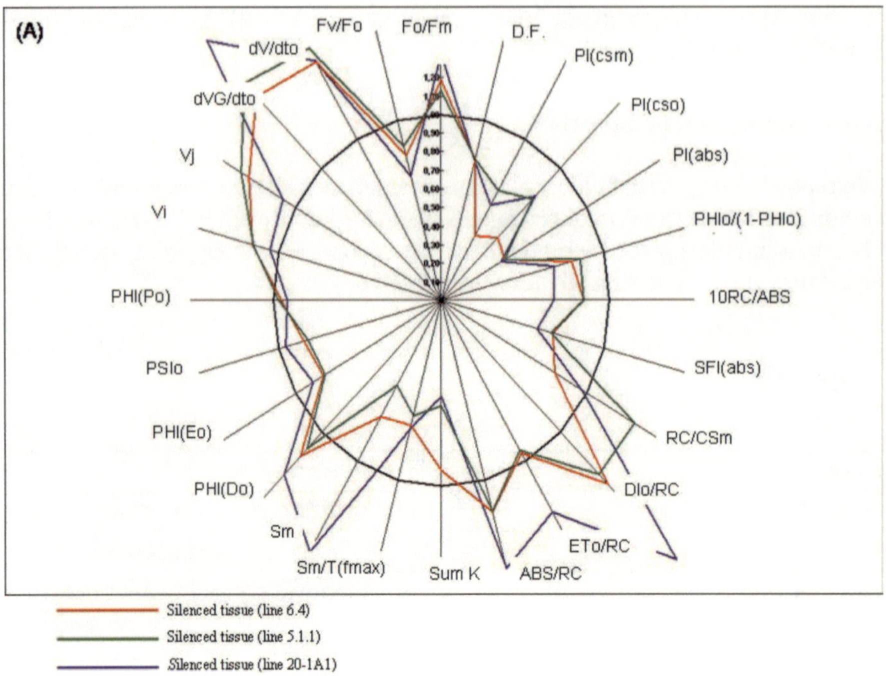

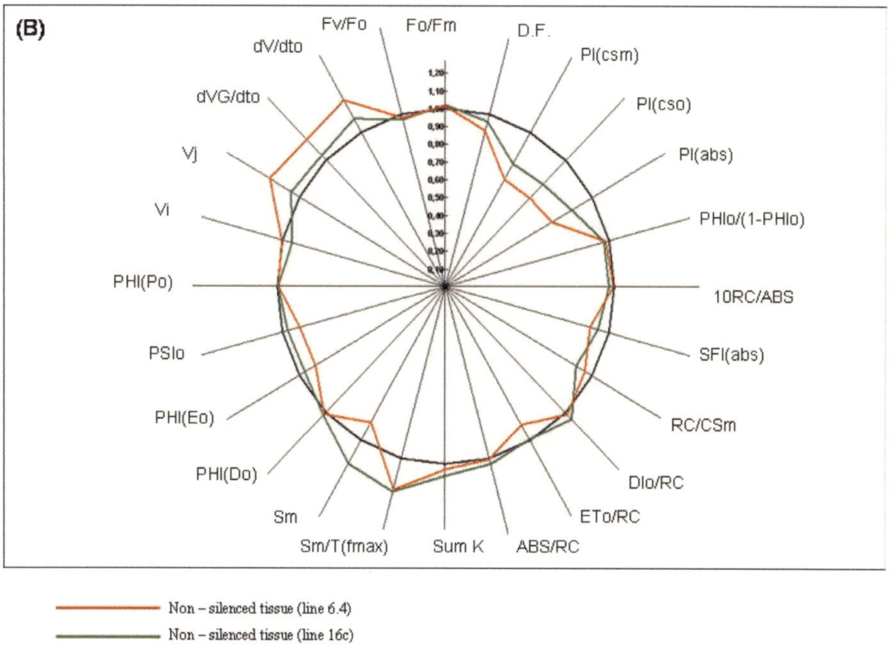

Figure A1.
A, B: Representative radar plots are denoting basic bioenergetic parameters of photosynthesis, in the examining tissues. The control condition (wt tissue) is presented as circle.

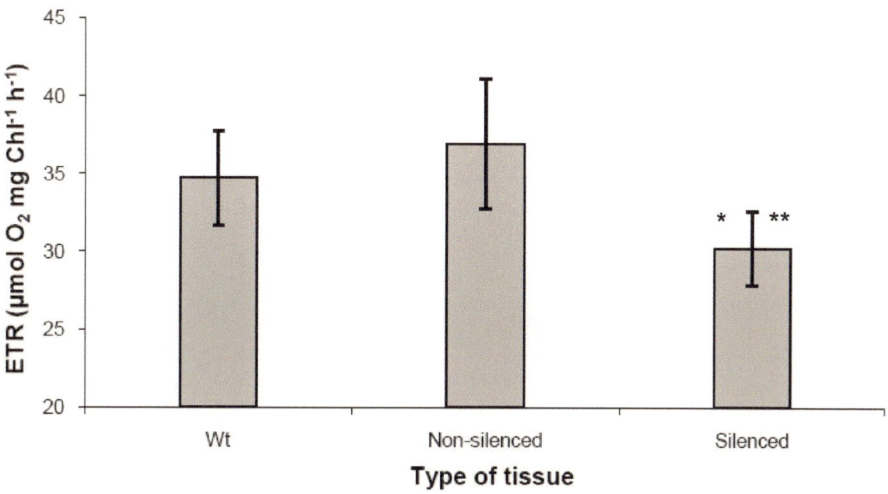

Figure A2.
In vitro *estimation of the linear electron flow in thylakoid membranes from the plant categories studied. Values normalized over wt ones. ETR: Electron Transport Rate, Chl: chlorophyll; h: hour.*

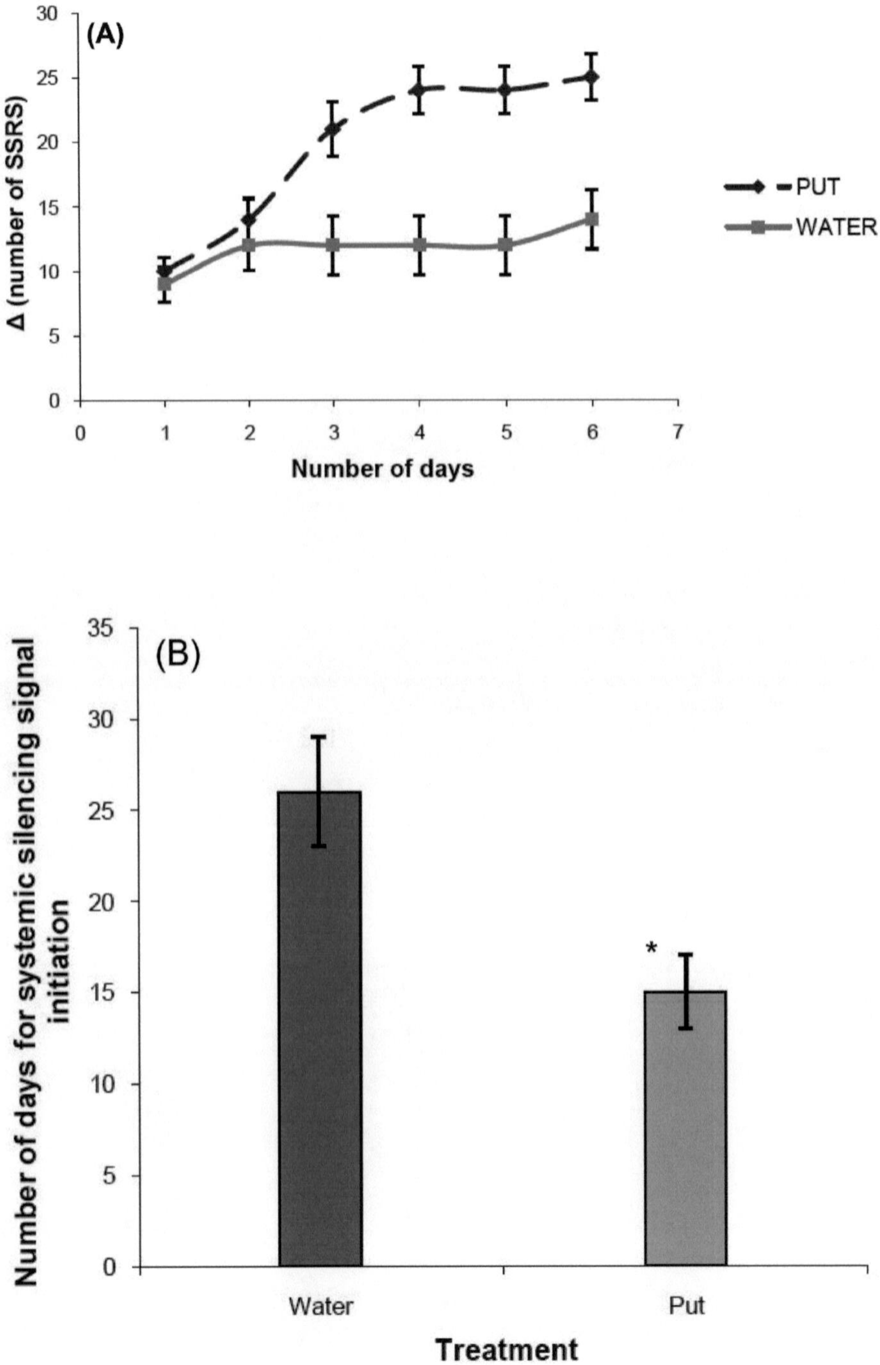

Figure A3.
A Daily monitoring for the number of short-range silencing spread (SSRS) events. B: Number of days for the systemic silencing signal onset in plants that express GFP (line 16c), after agroinfiltration with a hairpin GFP construct. Treatment of tissues studied with either 1 mM Put or water, was performed for 5 days, prior the beginning of the monitoring. The data presented in A are sum of SSRS events from the total surface of leaves per plant individual, examined for each treatment. The initial number of spots before the beginning of monitoring was considered and it was subtracted (Δ) from the new appearing ones during the time of monitoring. Similar method of sampling in the case of B, was followed.

Author details

Christos Kotakis
Citrusgenetics Cooperative Biotech Laboratory (citrusgenetics.org), Gastouni Ileias, Hellas

*Address all correspondence to: photoRNAchem@proton.me

References

[1] Vrettos N, Tournier B. The sound of silence in plants. Functional Plant Science and Biotechnology (Global Science Books). 2007;**2007**:94-99

[2] Kotakis C, Vrettos N, Kotsis D, Tsagris M, Kotzabasis K, Kalantidis K. Light intensity affects RNA silencing of a transgene in *Nicotiana benthamiana* plants. BMC Plant Biology. 2010;**10**:220

[3] Kotakis C, Vrettos N, Daskalaki MG, Kotzabasis K, Kalantidis K. DCL3 and DCL4 are likely involved in the light intensity-RNA silencing cross talk in *Nicotiana benthamiana*. Plant Signaling & Behavior. 2011;**6**(8)

[4] Koscianska E, Kalantidis K, Wypijewski K, Sadowski J, Tabler M. Analysis of RNA silencing in agroinfiltrated leaves of *Nicotiana benthamiana* and *Nicotiana tabacum*. Plant Molecular Biology. 2005;**59**:647-661

[5] Lichtenthaler HK, Buschmann C, Knapp M. How to correctly determine the different chlorophyll fluorescence parameters and the chlorophyll fluorescence decrease ratio R-Fd of leaves with the PAM fluorometer. Photosynthetica. 2005;**43**:379-393

[6] Strasser RJ, Tsimilli-Michael M, Srivastava A. Analysis of the chlorophyll a fluorescence transient. In: Govindjee Pa C, editor. Advances in Photosynthesis and Respiration. Chlorophyll A Fluorescence. A Signature of Photosynthesis. Dordrecht, The Netherlands: Springer; 2004. pp. 321-362

[7] Toth SZ, Schansker G, Strasser RJ. A non-invasive assay of the plastoquinone pool redox state based on the OJIP-transient. Photosynthesis Research. 2007;**93**:193-203

[8] Bellafiore S, Barneche F, Peltier G, Rochaix JD. State transitions and light adaptation require chloroplast thylakoid protein kinase STN7. Nature. 2005;**433**: 892-895

[9] Munne-Bosch S, Shikanai T, Asada K. Enhanced ferredoxin- dependent cyclic electron flow around photosystem I and -tocopherol quinone accumulation in water-stressed ndhB-inactivated tobacco mutants. Planta. 2005;**222**:502-511

[10] Casazza AP, Tarantino D, Soave C. Preparation and functional characterization of thylakoids from *Arabidopsis thaliana*. Photosynthesis Research. 2001;**68**:175-180

[11] Romanowska E, Drozak A, Pokorska B, Shiell BJ, Michalski WP. Organization and activity of photosystems in the mesophyll and bundle sheath chloroplasts of maize. Journal of Plant Physiology. 2006;**163**:607-618

[12] Andronis EA, Roubelakis-Angelakis KA. Short-term salinity stress in tobacco plants leads to the onset of animal-like PCD hallmarks in planta in contrast to long-term stress. Planta. 2010;**231**:437-448

[13] Navakoudis E, Vrentzou K, Kotzabasis K. A polyamine- and LHCII protease activity-based mechanism regulates the plasticity and adaptation status of the photosynthetic apparatus. Biochimica et Biophysica Acta. 2007; **1767**:261-271

[14] Breyton C, Nandha B, Johnson GN, Joliot P, Finazzi G. Redox modulation of cyclic electron flow around photosystem I in C3 plants. Biochemistry. 2006;**45**: 13465-13475

[15] Puthiyaveetil S, Ibrahim IM, Allen JF. Oxidation–reduction signalling

components and regulatory pathways of state transitions and photosystem stoichiometry adjustment in chloroplasts. Plant, Cell and Environment. 2011;**2011**:2349. DOI: 10.1111/j.1365-3040.2011.02349.x

[16] Wollman FA. State transitions reveal the dynamics and flexibility of the photosynthetic apparatus. The EMBO Journal. 2001;**20**:3623-3630

[17] Puthiyaveetil S. A mechanism for regulation of chloroplast LHC II kinase by plastoquinol and thioredoxin. FEBS Letters. 2011;**585**:1717-1721

[18] Pfannschmidt T, Nilsson A, Allen JF. Photosynthetic control of chloroplast gene expression. Nature. 1999;**397**: 625-628

[19] Johnson GN. Physiology of PSI cyclic electron transport in higher plants. Biochimica et Biophysica Acta. 2011; **1807**:384-389

[20] Ioannidis NE, Sfichi L, Kotzabasis K. Putrescine stimulates chemiosmotic ATP synthesis. Biochimica et Biophysica Acta. 2006;**1757**:821-828

[21] Qi Y, Denli AM, Hannon GJ. Biochemical specialization within Arabidopsis RNA silencing pathways. Molecular Cell. 2005;**19**: 421-428

[22] Tang G, Reinhart BJ, Bartel DP, Zamore PD. A biochemical framework for RNA silencing in plants. Genes & Development. 2003;**17**:49-63

[23] Matranga C, Zamore PD. Plant RNA interference in vitro. Cold Spring Harbor Symposia on Quantitative Biology. 2004; **69**:403-408

[24] Walters RG. Towards an understanding of photosynthetic acclimation. Journal of Experimental Botany. 2005;**56**:435-447

[25] Foyer CH, Shigeoka S. Understanding oxidative stress and antioxidant functions to enhance photosynthesis. Plant Physiology. 2011; **155**:93-100

Chapter 8

Whispers of Nourishment: Unveiling the Role of Non-Coding RNA in Plant Nutrient Availability

Jajati Keshari Nayak, Rashmi Chauhan, Sundip Kumar and Sharat Prabhakaran

Abstract

Micronutrients play a vital role in crop growth and development, as they are essential for various physiological processes and biochemical reactions within plants. Insufficient levels of micronutrients in the soil can lead to nutrient deficiencies in plants, resulting in stunted growth, reduced yields, and poor overall health. In the last decades, it has been seen that noncoding RNAs (ncRNAs) are involved in the regulation of micronutrient uptake, transport, and utilization in plants. They can modulate the expression of genes encoding transporters, chelators, and enzymes related to micronutrient acquisition and utilization. By fine-tuning gene expression, ncRNAs can help plants adapt to varying nutrient availability and optimize their nutrient uptake efficiency. Understanding the interplay between micronutrients and ncRNAs provides insights into the intricate molecular mechanisms underlying nutrient homeostasis and plant adaptation to nutrient stress. This study delves into the origin of various noncoding RNAs, such as miRNA, siRNA, and tsRNA, elucidating their pivotal role in maintaining micro- and macronutrient equilibrium within plant tissues. Overall, this research underscores the intricate interplay between micronutrients and noncoding RNAs in crop plants, shedding light on the intricacies of nutrient regulation and opening up new avenues for future investigation and potential applications in agriculture.

Keywords: noncoding RNA, miRNA, nutrient homeostasis, biofortification, micronutrient availability

1. Introduction

Plants rely on a minimum of 14 vital phytonutrients, primarily obtained from the soil, to support normal growth, development, and life cycle completion. Investigating the acquisition and distribution of these nutrients is crucial, as their metabolic and biochemical functions play a fundamental role in various parts of plant physiology and functional biology [1]. It is important to carefully study the tissue and subcellular concentrations of these nutrient elements within a limited physiological range [2]. Consequently, the uptake and assimilation of nutrients must be well-coordinated, considering the intricate and controlled interaction between mineral supply and plant demand. Despite the significance of these mineral nutrients, mostly are inaccessible for plants to uptake [1]. As a result, substantial efforts are being followed toward enhancing plant nutrient efficiency in diverse areas.

Plants acquire mineral ions from the soil, employing both long-distance and short-distance transport mechanisms to allocate them in different plant compartments. The distinctive expression patterns of various transporters, which adapt to mineral availability, intricately regulate the molecular processes orchestrating the absorption, conveyance, and loading of mineral ions into storage organs [3]. In instances of inadequate mineral ions within the soil, plants elicit responsive signaling cascades, where the dynamic interplay between signaling molecules and transporter or carrier genes contributes to an enhanced facilitation of ion movement [4]. When a plant encounters a nutrient deficiency, swift signaling molecules promptly relay this information to the plant's genetic material, ensuring the maintenance of nutritional equilibrium. The scarcity of nutrients triggers the activation of transporters or prompts alterations in root architecture and development, thus optimizing mineral absorption. This intricate process involves signaling molecules engaging in interactions with specific nucleotide sequences within messenger RNA (mRNA), thereby effecting modifications in gene expression [5].

Small regulatory RNAs, regarded as highly pervasive signaling molecules, wield significant influence over posttranscriptional gene expression modulation. These tiny RNA agents form precise bindings with mRNAs and engage in perfect base complementarities, which causes the attenuation of gene expression [6]. Small interfering RNAs (siRNAs) and micro-RNAs (miRNAs) are two of the most important types of small RNA in plant biology. They are closely linked to how genes are turned off. The former, which show up as small 20 to 24 nucleotide regulators, often come from endogenous genomic loci or are brought in from the outside as nucleic acid duplexes. A central role in this regulatory milieu is assumed by Dicer-like enzymes (referred to as DCL in *Arabidopsis*), adept at cleaving double-stranded duplexes, complete with hairpin loops, thereby yielding concise, impeccable duplexes. These are subsequently harnessed by the RNA-induced silencing complex (RISC). Within the RISC ensemble, the sense strand assumes the roles of recognition and binding to target genes or transcripts. This highly specific interaction leverages perfect nucleotide complementarities and consequently exerts regulatory control over gene expression, operating at either the posttranscriptional or transcriptional echelons [7].

Recent scientific inquiry has unveiled the presence of several novel miRNAs, a veritable frontier, which wield influence over the uptake and conveyance of nutrient minerals within the intricate web of plant biology [6]. The present comprehensive review delves into the contemporary understanding of diverse classes of nutrient-responsive miRNAs, delving deep into their functional roles within the intricate tapestry of nutrient homeostasis. This insightful exploration is poised to illuminate the role of miRNAs as paramount overseers, acting as maestros orchestrating the complex ballet of nutrient loading within plant systems. In this study, we tried to establish the interesting relation of different noncoding RNAs that might be able to control how transporter genes work. This opens the door to a new way of thinking that could have a lot of effects. This paradigm has the potential to revolutionize the landscape of crop biofortification, a process aimed at enhancing the nutritional profile of crops and thereby unraveling new vistas of mineral bioavailability within cereal grains, thereby elevating their intrinsic nutritional value. This, in turn, holds the promise of reshaping the future of agriculture and human nutrition.

2. Biogenesis of noncoding RNAs and its classifications

Various hypotheses are associated with the origins of noncoding RNAs (ncRNAs), exhibiting varieties in biological roles [8]. miRNAs, siRNAs, piRNAs, and lncRNAs are the main types of ncRNAs. These ncRNAs are generated through different mechanisms like duplication of genes, transcription of pseudogenes, from hairpin structure of double-stranded RNA, from heterochromatin regions of DNA and through transposable elements [9].

2.1 tsRNA

tsRNA (transfer RNA-derived small RNAs) is a class of ncRNA; 18–40 nucleotides long fragments originate from tRNA and can undergo site-specific cleavage by various endoribonuclease like Elac Ribonuclease Z 2 (ELAC2)/RNase Z, angiogenin, Dicer, and RNase L, which represent the variety pertaining to their sequence, length, and functional properties. 3'URF, 5′ or 3′ Trh, and 3′ or inter-tRFs are three main categories of tsRNA [10]. Majorly known tsRNA predominantly arise from tRNA, yet the exact mechanism of the biogenesis of tsRNA biogenesis remains uncertain. Nevertheless, some hypotheses have been suggested based on research. It is suggested that different enzymes like *Arabidopsis* S-like Ribonuclease 1 (RNS1)/DCL1 (Dicer-like protein) [11] take part in the synthesis of mature tsRNAs. tsRNA's role has been validated in facilitating the acclimatization of plants to adverse environmental condition like stress by regulating target genes during both transcriptional and post-transcriptional phases.

2.2 miRNA

miRNAs, a significant class of ncRNA, are molecules of 20–22 nucleotides in length (**Figure 1**) [13]. Synthesis of miRNA begins with the miRNA gene, which transcribed product is processed with the assistance of RNA polymerase II followed by RNase cleaves pri-miRNA hairpin to yield a precursor-mi RNA (pre-miRNA). Subsequently, another cleavage event releases an miRNA/miRNA* duplex (**Figure 2**) [15]. Nuclear HUA ENHANCER 1 (HEN1) protein stabilizes miRNA complex (miRNA/miRNA*) through methylation process in its 3' prime region [16]. The next step begins with ARGONAUTE 1 (AGO1) protein, which combines with processed miRNA in nucleus and transports into cytoplasm, where it function as posttranscriptional gene silencing [13, 17]. A diverse range of studies in different aspects of plant reveal that miRNA governs the fundamental developmental process like flowering time, leaf morphogenesis, as well as stress conditions such as biotic and abiotic stress [18].

2.3 siRNA

Based on the mechanism of action and biogenesis, siRNA can be classified into three subtypes including heterochromatic siRNAs (hc-siRNAs), natural antisense siRNAs (nat-siRNA), and transacting siRNA (tasiRNAs). tasiRNA operates in a cis manner, distinct from its biogenesis site, and exhibits similar regulatory mechanism like miRNA, while hc-siRNA acts effectively in a cis manner within its origin or the region that is homologous to its origin [19]. nat-siRNA is further grouped into trans-nat-siRNAs and cis-nat-siRNA. Cis-nat-siRNA originates from the transcription

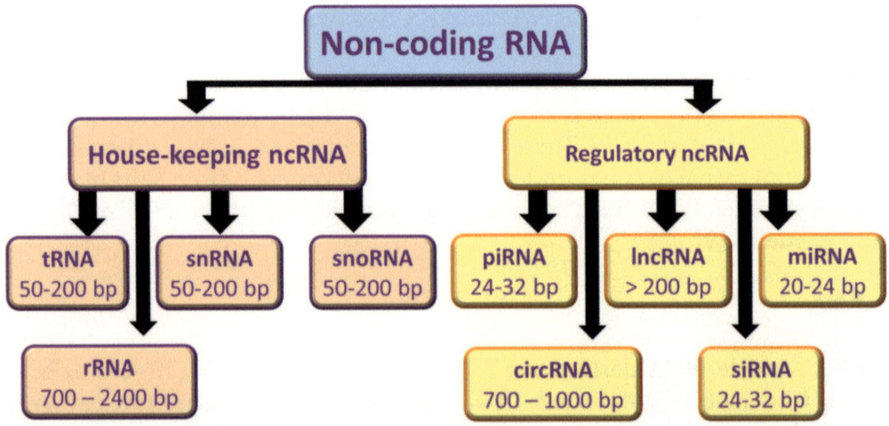

Figure 1.
Flowchart demonstrating the classification of noncoding RNA (nc-RNA) and their length in nucleotides. Figure inspired from [12].

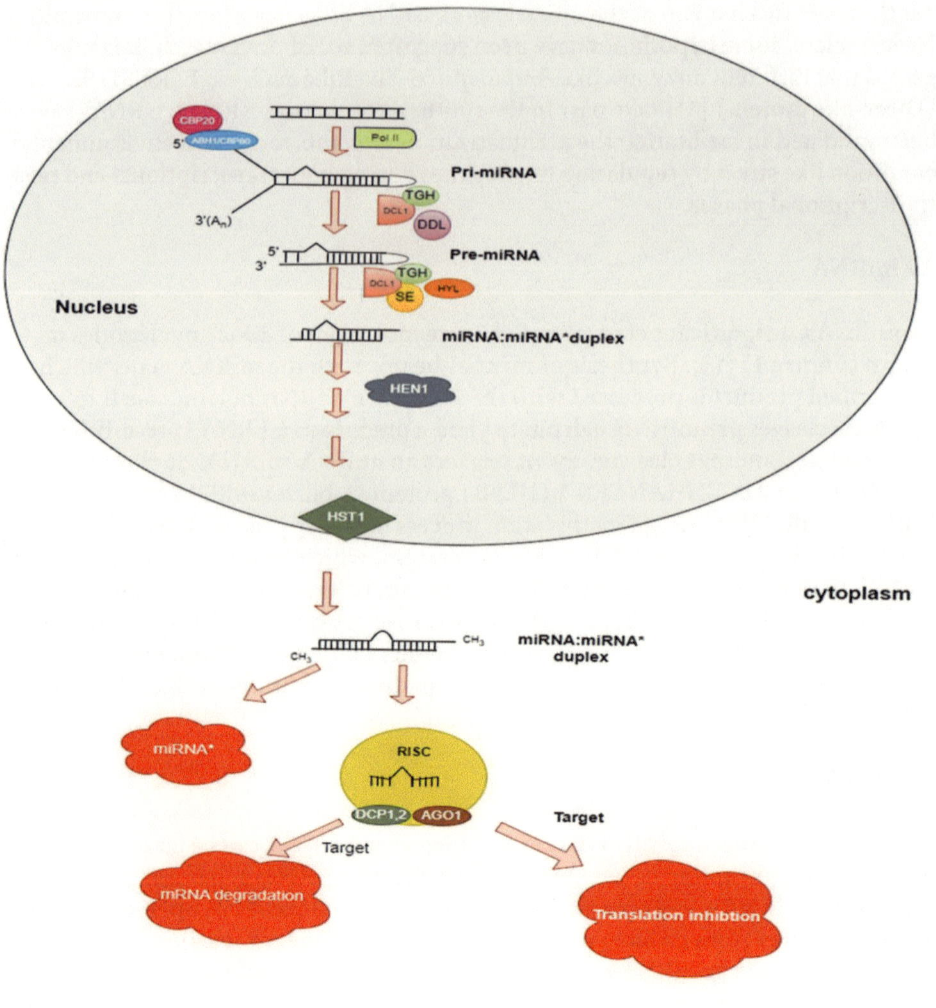

Figure 2.
Flowchart of miRNA biogenesis. Figure inspired from [14].

of opposite strands of the same genomic location, while trans-nat-siRNA from the transcription of multiple genomic locations [20]. Synthesis of siRNA involves dsRNA, RNA-induced silencing complex with AGO protein, and DCL enzyme. RNA-induced silencing complex (RISC) governs the regulation at the posttranscriptional or transcription level [21]. DCL4, RNA-dependent RNA polymerase 6 (RDR6), and suppressor of gene silencing3 (SGS3) are proteins involved in the synthesis of tasiRNA [22], while nat-siRNAs synthesis depends on the activities of RDR6, DCL2, SGS3, DCL1, and NRPD1A (plant-specific RNA polymerase) [23]. DCL3, 5 s rRNA, Pol IV, NRPD1A, and RDR2 are involved in the production of hc-siRNAs [24]. siRNA's role has been established in the developmental process, defense against bacteria, eukaryotic pathogen, and plant growth [25].

2.4 lncRNAs

lncRNAs (long noncoding RNAs) are similar to mRNAs like having poly A tail at the 3′ end and m^7G cap at the 5′ end but have differences in length and exon number. lncRNA can be categorized into five distinct groups based on different transcription site by pol II, pol III, polIV, or polV. (1) sense lncRNAs are transcribed on the same strand as exons; (2) antisense lncRNAs are transcribed on the opposite strand of exons; (3) intronic lncRNAs are transcribed within introns; (4) intergenic lncRNAs are located between two separate genes; (5) enhancer lncRNAs originate from an enhancer region associated with protein-coding genes [12]. These lncRNAs exert control over target regulation through various mechanisms like chromatin remodeling [26], transcriptional repression, RNA splicing, as well as transcriptional enhancement [27]. Several studies support lncRNA's important role in flowering time, abiotic stress, and reproduction [28]. Interestingly, lncRNAs can encode small peptides that are also important for numerous cellular process [29].

3. Role of nc-RNAs in soil: Plant-nutrients homeostasis

3.1 miRNAs in nitrogen (N) homeostasis

Among the array of macronutrients, nitrogen assumes a crucial role in the formation of vital cellular constituents like proteins, nucleic acids, and chlorophyll. This significance makes it indispensable for the comprehensive growth of plants [30]. Plants acquire nitrogen in various molecular forms from soil, encompassing ammonium, nitrate, and urea. Notably, nitrate (NO_3^-) stands as the preferred form among these three, which subsequently undergoes for conversion into nitrite in plant cell and then into ammonium. These converting steps are facilitated by enzymes and culminate in the synthesis of amino acids [31]. So, scarcity of nitrogen in plants places the plant in a precarious situation. There is a wide range of approaches by which plants can overcome the nitrogen-limiting condition. But among all approaches, miRNAs are the efficient tools by which a plant changes internally at transcriptional and posttranscriptional levels to survive in harsh conditions.

The transcriptional patterns of diverse miRNA families have been detected in distinct agricultural plant species in the context of nitrogen. Examples includes soyabean, *Arabidopsis thaliana, Bouteloua gracilis* [32–40]. A comprehensive summary of various miRNAs participating in the response to nitrogen limitation and their target is listed in **Table 1**.

Sl. No.	miRNAs family	Target	Function	Plant species	Reference
1.	↑miR167	↓AUXIN RESPONSE FACTOR 8	Trigger lateral root outgrowth	Maize	[41]
2.	↑miR160 and miR171	↓ARF10/16/17	Lateral and adventitious root development	Maize	[42]
3.	miR171	SCARECROW-LIKE PROTEIN 6	Root development	*Arabidopsis*	[43]
4.	miR393	AUXIN-SIGNALING F-BOX PROTEIN 3 (AFB3)	Root system architecture	Maize	[44, 45]
5.	↓miRNA169	↑NUCLEAR FACTOR Y SUBUNIT A5 (NFYA5)	Downregulates N uptake	Maize	[46]
6.	↓miR164	↑*NAC1*	Lateral root production	Arabidopsis	[47]

↓ *down regulation* ↑ *upregulation.*

Table 1.
Function and target gene of miRNA for nitrogen homeostasis in different plants.

3.1.1 NcRNAs orchestrating nodulation and symbiosis in leguminous plant for nitrogen uptake

Higher plants demonstrate a range of adaptive strategies to enhance the capacity of roots to uptake nitrogen from soil. Among all forms of nitrogen available for uptake, ammonium is harnessed through the process of symbiotic nitrogen fixation. In this, ammonium is culminated *via* bacterial enzyme nitrogenase followed by conversion into accessible form [48]. Before the commencement of N_2 fixation, nodule development in the pulse crop follows different steps in succession starting from nodule organogenesis, bacterial infection, and the initiation of nitrogen fixation [49]. This process is triggered by the exchange of diffusible chemical signals between the host plant and symbiotic partner followed by bacterial penetration *via* infection into plant's root hairs. Origination of nodule primordium is the attribute of infection thread, which divides into the plant cortex. Eventually, bacteria are liberated from the infection thread by endocytotic mechanism, leading to the formation of symbiosomes, where nitrogen fixation is executed [50]. In the favor of fixing nitrogen, bacteria receive carbohydrates as well as mineral nutrients.

Nodule development process is fully investigated and characterized at the molecular level [51]. On the other hand, miRNA also has a role in the nitrogen fixation process. Many miRNAs have been discovered till date who are involved in legume-rhizobia symbiosis [52]. Lelandais-Brière et al. [53] identified 100 conserved and novel miRNAs from nodules synthesized in *Medicago truncatula* roots 21–30 d postinoculation and suggested their role in the development of nodule. In soybean (*Glycine max*), sRNA library is sequenced and total 129 miRNAs generated in which 87 miRNA were novel from distinct developmental stages like root, seed, flower, and nodules [54]. In this pulse crop 15 small RNA libraries were sequenced in young, mature, and senescent nodules stages and 139 miRNAs were identified. Five parallel analysis of

S. No.	miRNA	Target	Function	Crop Species	Reference
1.	miR166	HD-ZIP III TF	Root nodule development inhibition	*Medicago truncatula*, *Glycine max*	[55, 56]
2.	miR171c	NSP2 TF	Nodule infection	Legume lotus japonicas *M. truncatula*	[57]
3	gma-miR171o, gma-miR171q	GmSSCL-6, GmNSP2	*NIN,ENOD40* and *ERN* genes	*Glycine max*	[58]
4	miR396b	MtNSP2	Nodule formation	*Medicago truncatula*	[59–61]
5	miR393j-3p	Early Nodulin 93	Inhibition of nodule formation	*Glycine max*	[62]
6	miRNA169	Glyma10g10240 and Glyma17g05920	Inhibit HAP proteins	Soybean (*Glycine max*)	[63]
7	miR2118, miR2109, and miR1507	silencing of target NB-LRR genes post-transcriptional	Symbiotic interaction with rhizobia	*Medicago truncatula*	[64]

Table 2.
Role of different miRNAs in nodulation of leguminous crops for nitrogen uptake.

RNA ends (PARE) libraries sequenced and identified 533 miRNA targets including 8 nodule-specific genes. In **Table 2**, we have provided the function and list of miRNAs in the nodulation process.

3.2 Role of miRNA in phosphorus (P) homeostasis

Phosphorus (P) stands as a pivotal element supporting an array of physiological and biochemical functions. It actively engages in a diverse spectrum of metabolic processes, notably in the synthesis of nucleic acids and the generation of energy within plants. It enters a wide range of metabolic processes, specifically, the synthesis of nucleic acids and energy generation of plants, which makes it hard for plants to grow under P starvation. This significance renders it challenging for plants to thrive in conditions of phosphorus scarcity.

Recent investigations have shed light on the role of miRNAs (small RNA molecules) in orchestrating the intricate regulation of phosphate-related gene expression across various plant species, such as *Arabidopsis*, rice, wheat, barley, maize, soybean, white lupin, and tomato [65]. In *Arabidopsis*, the detection of low phosphate levels by the shoots promptly triggers the synthesis of miR399a-f families. These miRNAs, particularly miR399, intricately bind to complementary sites within the phosphate over accumulator 2 (PHO2) transcript, positioned in the 5′ untranslated region (UTR) about 200–400 base pairs upstream. This binding event triggers mRNA cleavage. Notably, PHO2 serves as a pivotal transporter facilitating phosphate mobilization and is linked to the E2 ubiquitin conjugating enzyme. The downregulation of PHO2 assists in accumulating substantial phosphate levels within the shoots. Conversely, a long noncoding RNA known as induced by phosphate starvation1 (IPS1) plays a counteractive role. IPS1 inhibits the miR399-mediated regulation of the PHO2 gene, safeguarding the PHO2 transcript from degradation [66].

Remarkably, miR398a's expression is subject to regulation not only by phosphorus availability but also by carbon (C) and nitrogen (N) limitations, hinting at its broader involvement in mineral homeostasis. Suppression due to carbon limitation has also been shown to contrast with its induction by sucrose [67]. Additionally, in the context of soybean roots, during phosphorus deficiency, miR159a experiences upregulation, while miR319a, miR396a, miR398b, and miR1507a undergo downregulation. This involves the modulation of miRNA expression in response to phosphorus starvation hinges on the prevalence of phosphorus-responsive motifs (P-responsive motifs) within the miRNA gene promoters. Notably, the cis-acting regions of miRNA genes are reported to contain a higher abundance of P-responsive motifs compared to their nonresponsive counterparts [68].

3.3 Role of miRNA in potassium homeostasis

Potassium (K) is a critical inorganic nutrient that is essential for plant growth, development, and yield formation in cereal crops [69]. Many agricultural soils around the world are K deficient, which limits sustainable crop development [70]. In *Arabidopsis*, the first upregulated miRNA, miR156, was characterized in the K deficiency condition. Later on, other miRNAs like miR169, miR395, and miR398 were also found to be responsive under K deficient conditions [71]. These miRNAs target genes like *MADS-23*, *MADS-27a*, *MADS-27b*, and *MADS-57*; synthesize a transcription factor; and ultimately endow the plant with nutrient deficient conditions [41]. Previous research has shown that the module miR319/TCP4 and the pathways facilitated by the module miR396/GRF play a significant role in enhancing barley's ability to withstand low-K stress. Additionally, it has been found that ata-miR1432-5p functions as a regulator in the transduction pathways of Ca2+ signaling, which are triggered by low-K stress [42]. Furthermore, the involvement of stu-miR530_L-2R + 2 in potato has been shown in the modulation of responses to K treatments *via* the regulation of a specific gene that encodes a protein belonging to the Zinc Knuckle (CCHC-type) family, known as protein [43]. In barley, it has been reported that miR408 responds to low K through the blue copper protein (BCP) [44]. Collectively, these data suggest that microRNAs (miRNAs) have a role in facilitating plant responses to nutritional stress.

In wheat, an insightful exploration by Zhao [45] unearthed a noteworthy revelation. Among the 89 miRNA families investigated, a distinctive pattern of differential expression emerged under conditions of potassium (K) deficiency. This encompassed 68 families with known attributes, while an additional 21 families were newly identified. Notably, 11 specific miRNAs, namely, novel_17, miRNA319, miRNA531, miRNA9773, miRNA9670-3p, miRNA398, miRNA159a, miRNA9778, miRNA408, miRNA9776, and miRNA1133, demonstrated close associations with plant resilience in the face of K starvation. Delving deeper into these differentially expressed miRNAs through gene ontology and KEGG analysis, their roles were found to fall within the purview of ADP-binding activity and secondary metabolic pathways. A notable instance was the robust impact of tae-miR408 over expression, which not only enhanced plant morphology but also bolstered K-deficiency tolerance. This enhancement was achieved by elevating K uptake efficiency, stimulating photosynthetic pigment biosynthesis, and fostering a harmonious balance in ROS homeostasis.

3.4 Role of miRNA in calcium (Ca) homeostasis

Calcium, a fundamental macronutrient indispensable for plants, is required in significant quantities. Its pivotal role encompasses shaping plant cell structures and occupying a central position as the predominant secondary messenger in relaying responses to environmental stresses. According to some studies, cation channels and the apoplast facilitate the uptake of Ca^{2+} by plants from the soil. Following this absorption, the transpiration stream facilitates the subsequent transport of Ca^{2+} to the shoot through the xylem [46].

MicroRNAs (miRNAs) have significant influence across various dimensions of plant biology, spanning growth, development, stress responses, and epigenetic regulation. In a recent study on the orphan crop *Eragrostis tef*, researchers have found a strong link between the amount of miRNA in the root and shoot tissues and how much calcium (Ca) the plant took in [72]. This investigation unveiled noteworthy insights like *E. tef* plants enduring prolonged Ca^{2+} deficiency exhibited the identification of 1380 miRNAs, whereas control plants manifested the detection of 1495 miRNAs. In this comprehensive endeavor, a total of 509 miRNAs surfaced, comprising 161 miRNAs akin to those characterized in diverse plant species and 348 novel miRNAs. The remaining miRNAs remained enigmatic in their characterization. Furthermore, comprehensive predictions were made for the putative target genes and their roles, corresponding to both the known and newly identified miRNAs. Through gene ontology (GO) analyses, a diverse array of biological and molecular functions surfaced for the projected target genes, including roles in calcium absorption and transport. The discernment of differentially expressed miRNAs underscored distinct enrichments in either root or shoot tissues of plants exposed to low Ca^{2+} conditions. The potential of delving deeper into the intricacies of these miRNAs and their intertwined targets from this research venture stands to offer a heightened comprehension of how *tef* and similar underexplored crops respond to calcium deficiency.

In parallel, this paradigm extends to the area of peanut (*Arachis hypogaea*), warranting comparable in-depth exploration. A set of twelve distinct miRNA families specific to peanuts were identified, encompassing a group of 29 previously recognized miRNAs and 132 potential novel ones. According to the research by Chen [73], a total of 87 miRNAs exhibited varying expression patterns during the initial stages of embryonic development, both in conditions of insufficient calcium and at adequate calcium levels. The study further pinpointed 117 target genes that displayed differing levels of expression. An integrated investigation combining miRNA and transcriptome expression revealed that twenty miRNAs underwent expression changes, consequently impacting the expression of 52 target genes. By comparing gene chip analysis and transcriptome sequencing, unique targets with distinct expression patterns were identified. The study's findings underscore the substantial role of microRNAs (miRNAs) in governing the expression of genes implicated in embryonic development. In particular, genes such as TCP3, AP2, EMB2750, GRFs, and cytochrome P450 (CYP707A1 and CYP707A3) involved in abscisic acid (ABA) signaling, along with BR1, which facilitates brassinosteroid (BR) transport, are actively influenced by miRNAs. Both miRNAs and their corresponding target genes are thought to participate in the process of peanut embryo abortion as a response to inadequate calcium. This study reveals the presence of miRNA-mediated regulatory systems involved in embryo abortion in peanuts, particularly when calcium is lacking [73].

3.5 Role of miRNA in magnesium (Mg) homeostasis

Magnesium (Mg), a core macronutrient involved in plant growth and development, has pivotal function in chlorophyll biosynthesis and acts as cofactor in carbon metabolism [55]. Magnesium deficiency conditions are prevalent in sandy, acidic soils where magnesium is very susceptible to leaching. Although the importance of magnesium in plants is well established, miRNA's involvement in Mg homeostasis is relatively underexplored. Liang [56] conducted a study in *Citrus sinensis* to identify miRNAs that provide tolerance in Mg-deficiency conditions *via* Illumina sequencing. They identified 101 miRNAs were upregulated and 69 miRNAs downregulated in roots. In Mg-deprived conditions, miR158 was expressed, which repressed its target gene *SPFH* (stomatins, prohibitins, flotillins, and HflK/C). In Mg-starved conditions, miR1044, miR5029, miR6485, miR6190, and miR3437 control the expression of transport-related genes. Root development controlled by miR5261, miR6485, and miR158 in magnesium reduces conditions [56]. miR2919 inhibits root respiration by targeting phosphoenol pyruvate carboxylase 3 (PEPC3), which eventually induces low respiration and reduced accumulation [57]. Surprisingly, one more interesting finding suggested that disease resistance capacity elevated in Mg deprives plants by targeting *NB-ARC domain-containing disease resistance protein* through miR780 and miR6278 [56].

3.6 Role of miRNA in Sulfate (SO4) homeostasis

Sulfur, a crucial micronutrient for plant, is available in the form of carbohydrate, proteins, and sulfo-lipid. Inorganic sulfur is the most predominant form of sulfur that is translocated into different tissue. SULTR1;1 (sulfate transporter1;1) and SULTR1;2 (Sulfate transporter1;2), two vital S carriers, are responsible for sulfate uptake from topsoil [58]. SLIM1, a transcription factor, regulates sulfur utilization by controlling transporter like SULTR1;2 [59]; the enzyme APS, found in various isoforms enzyme (APS1, APS3, APS4), is a key player in S assimilation pathway [6]. APS1, APS4 within plastid facilitate sulfur reduction, leading to cysteine formation [60]. Sulfate-deprived condition triggers to activate *SULTR1-2,* SLIM1, and other S-deprivation response genes, which eventually upregulates miR395. Up regulation of miR395 lead to downregulation of three APS genes (APS1, APS3, APS4) and SULTR2;1 transcripts which eventually accumulate the sulfate in shoot and disrupt the transport of sulfate from mature leaves to young leaves respectively [61]. This cascade of pathways ultimately affects S uptake, facilitates phloem transport from roots to shoots, and also concurrently inhibits xylem-mediated shoot-to-root transfer [62]. Sulfur-deprived conditions change the expression of miR167, miR168, miR167, miR394, miR164, miR160, and miR156 in *Brassica napus* [63].

3.7 Role of miRNA in iron (Fe) homeostasis

Iron is an essential micronutrient, serves as cofactor in the electron-transport chains of both photosynthesis and respiration, and also functions in other cellular processes. Low availability of iron initiates a condition of stress in plants, as well as its high accumulation exerts detrimental impacts on plants by disruption lipid, protein, and DNA through Fenton catalytic reaction. miRNA responds in both situations, iron deficiency and iron toxicity. In an experiment, 8 miRNAs were cloned and small

RNA library was constructed from *Arabidopsis* seedlings under Fe-deficiency condition, and it concluded that the expression of miR394b, miR173, miR172c, miR172d, miR169b, and miR169c families initially was higher followed by a lower expression, but miR394a and miR159a were upregulated at the end of trial during Fe-deficiency period [64].

Paul [74] proposed that significant difference in iron content between transgenic plants and their WT (wild type) counterpart was documented during the milk stage in an experiment, resulting in upregulation of some transporters in roots of transgenic plants by over-expressing soyFER1 gene. They also found a reciprocal association between low miRNA expression, specifically miR9, miR38, miR21, and miR28, and higher expression of target *SAM-dependent carboxyl methyl transferase* gene in root tissue. Notably, the study unveiled that expression level of OsNRAMP4 (iron transporter) was elevated in roots during the milk stage with lower expression of the four novel miRNAs: miR11, miR26, miR30, and miR31.In one study, miR399 and miR408 families were found to be significantly expressed in transgenic plant roots that enhance the expression of iron transporters by minimizing the activity of a repressor gene [75].

3.8 Role of miRNA in zinc (Zn) homeostasis

Zinc stands out as a highly essential micronutrient that serves a pivotal role as a catalytic, regulatory, or structural cofactor for a multitude of enzymes and regulatory proteins within both plant and animal systems [76]. The exploration of distinct miRNAs has yielded valuable insights into the intricate process of zinc ion uptake from soil. Through the utilization of miRNA microarrays, research focused on *Sorghum bicolor* has pinpointed eight miRNA families that exhibit responsiveness to zinc deficiency. Particularly, miR398 and miR528 were found to target two Cu/Zn superoxide dismutase genes, *SbCSD1* and *SbCSD2*, respectively [77].

Numerous zinc-responsive microRNAs (miRNAs) play pivotal roles in regulating plant growth and development. MiR408 targets plantacyanin, influencing ROS signaling, defense, and reproduction. MiR166 regulates root development through HD ZIP proteins. MiR168 affects AGO1, while miR171 targets the SCL6 transcription factor. MiR319b modulates jasmonic acid production, and miR528 controls CSD expression [77]. Among them, miR171 and miR528 showed significant downregulation in leaves, while miR398, miR319, miR166, miR168, and miR399 were upregulated in roots. Resupplementation experiments validated their roles in phytohormone biosynthesis. Zinc deficiency influenced CSD expression differently for miR398 (downregulation) and miR528 (upregulation) [77].

In a similar vein, according to SHI [78] report, *Brassica juncea* has also demonstrated a variety of miRNAs that control zinc uptake. Within the spectrum of 21 conserved miRNA families, the spotlight falls on nine families displaying differential expression patterns. When *B. juncea* roots did not have enough Zn, 13 miRNAs were found to be upregulated, while miR399b and miR845a were downregulated. When the roots of *B. juncea* are exposed to abiotic stress, these miRNAs are changed. This changes how phytohormones react, how plants grow, and how they react to abiotic stress. In the case of rice, a full combination of miRNAome and transcriptome data showed that 12 different genes had different levels of expression and could be targets for 10 miRNAs that respond to Zn. Among these miRNAs are miR171g-5p, miR397b-5p, miR398a-5p, and miR528-5p [44].

3.9 Role of miRNA in manganese (Mn) homeostasis

Manganese (Mn) plays a significant role as a micronutrient in overseeing the progression and maturation of plants. It serves as the metallic element within several proteins and enzymes, including superoxide dismutase (SOD) and the oxygen-evolving complex within photosystem II. Furthermore, Mn plays a role in the creation of secondary metabolites [79]. It is an inorganic catalyst, facilitating the enzymatic processes necessary for plant growth. Mn plays a crucial role in various biochemical functions of plants, including photosynthesis, respiration, nitrogen assimilation, and pollen germination. It is indispensable for pollen tube growth, elongation of root cells, and shielding roots from pathogens [80]. However, limited research has delved into Mn toxicity in plants, necessitating further investigations into the potential modulation of Mn homeostasis by miRNAs and their target genes. As a result, a comprehensive understanding of posttranscriptional regulation regarding Mn homeostasis remains challenging to obtain.

Manganese (Mn) homeostasis has been observed to trigger changes in the expression of certain microRNAs (miRNAs). Notably, an excess of Mn leads to an increase in the expression of miR826. This miRNA specifically targets AOP2 and its related counterparts, AOP1 and AOP3. These genes, which are called AOPs as a group, make enzymes called 2-oxoglutarate-dependent dioxygenases. These enzymes help to make glucosinolates. It is reported that when miR826 is over-expressed in transgenic plants, the levels of glucosinolates and anthocyanins go down and the plants can handle less nitrogen (N) better. These findings suggest that excessive Mn disrupts nutrient balance by influencing the expression of miRNAs. Another miRNA, miR781, is associated with mini chromosome Maintenance 2 (MCM2), a gene known to influence root meristem function and play an essential role in embryonic development, as indicated by research by Sabelli [81]. The upregulation of miR781a and miR781b in Mn-treated seedlings corresponds to observable changes in root growth patterns. Moreover, miR5595 and miR5995b are predicted to target Methyl Esterase 7 (MES7) and MES9, genes involved in the hydrolysis of methyl-SA (MeSA) and systemic acquired resistance in plants [82]. Intriguingly, both miR5595a and miR5995b are upregulated in response to Mn treatment. These collective findings highlight the intricate relationship between miRNA expression and the effects of Mn homeostasis on various aspects of plant physiology and growth.

Researchers employed both miRNA microarray hybridization and qRT-PCR techniques to identify miRNAs responsive to Mn stress in common beans (*P. vulgaris*). The study revealed 37 miRNAs exhibiting distinct expression patterns under abiotic conditions and Mn stress. Notably, Mn toxicity induced the activation of 11 miRNAs while inhibiting 11 others. New Mn-responsive miRNAs, including miR1508, miR1515, miR1510/2110, and miR1532, were discovered. Key targets among the Mn-responsive miRNAs were found to be leucine-rich repeat-resistant proteins, receptor kinase proteins, and calcium-dependent protein kinases [83]. Furthermore, recent research focusing on *Arabidopsis* has demonstrated differential expression of miRNAs linked to cell growth, nutrient balance, and ion transport under Mn stress conditions [84].

3.10 miRNA in boron stress

Boron (B) plays a vital role in promoting proper plant growth and development by engaging in a range of physiological processes. These include supporting protein and nucleic acid metabolism, metabolizing lipids, facilitating cell division, and maintaining cell walls [85]. Research indicates that miRNAs contribute to responses against B-induced stress. For instance, barley's B-stress responsive miRNA network and associated pathways have been investigated [86]. Moreover, a comparative study of miRNA profiles between root and leaf samples was conducted. Within barley, 31 established and three novel miRNAs were identified, with 25 of them showcasing responsiveness to B treatment. Specific miRNA expression patterns were observed in distinct tissues; for example, miR156, miR171, miR397, and miR444 were exclusively expressed in leaves. These miRNAs specifically targeted and degraded 934 transcripts in barley. Furthermore, in the context of elevated B levels, variations in miRNA expression were detected in French beans (*Phaseolus vulgaris L.*), and the impacts of miRNA targets on gene expression under B stress were explored. This investigation was corroborated by Gene Ontology (GO) analysis, confirming the involvement of plant miRNAs in various cellular processes, including the circadian rhythm and vegetative development [87].

Citrus trees frequently encounter a widespread issue of boron (B) deficiency. Several studies have revealed the presence of B-responsive miRNAs in citrus species. Through high-throughput Illumina sequencing (HTIS) of *C. sinensis* roots, it was uncovered that 52 miRNAs were upregulated and 82 were downregulated. This showcases the remarkable adaptability of roots, which aids plants in enduring B-depleted conditions [88]. Certain miRNAs were proposed to influence root adaptation to B deficiency in several ways: (a) enhancing miR474 while reducing miR782 and miR843 to counteract reactive oxygen species (ROS); (b) boosting miR394 expression while dampening miR5023 expression to heighten cellular sensitivity to B deficiency; (c) lowering miR830, miR5266, and miR3465 levels to enhance fluid transport within cells; (d) managing osmo-protection *via* miR474 and other metabolic reactions through miR5023 and miR821. The research revealed that other miRNAs, such as miR472 and miR2118, exhibited reduced expression as B deficiency progressed, subsequently diminishing the presence of disease resistance genes and reducing root disease resistance [88]. Likewise, an examination of trifoliate orange (*Poncirus trifoliata*) through RNA transcript analysis revealed a decrease in miR397 levels following excessive B treatment. This treatment led to elevated transcription of laccase7 (LAC7), a target of miR397. Moreover, the treatment notably heightened laccase activity, underscoring the pivotal role of LAC7 in protein biosynthesis [89].

It has been shown that too much boron (B) has a genetic effect on wheat, especially when it comes to the expression of jasmonate (JA), ethylene, and changes in the cell wall. Under hazardous B conditions, *Arabidopsis* miRNAs that target JA and ethylene-related transcription factors (miR172 and miR319) and laccase (miR397) were studied. For the analysis of mature miRNA expression, Kayihan [90] recommended stem-loop qRT-PCR. In response to mild B toxicity, high levels of miRNAs that target transcription factors involved in the metabolism of jasmonic acid (JA) and ethylene stood out. However, this response was notably absent under severe B-toxicity conditions (designated as condition 3B). Among the regulating factors, the predominant influence was attributed to the miR172 and miR319 genes. On the other hand, the expression of miR397 exhibited no statistically significant alteration upon exposure to B toxicity, indicating that laccase-induced changes in the cell wall are not subject to posttranscriptional control. Furthermore, *Arabidopsis thaliana* exhibits a multifaceted ability to perceive oxidative stress and counter boron toxicity. This is achieved through the activation of miRNA expression, specifically targeting transcription factors associated with JA and ethylene metabolism [90].

3.11 miRNA and copper stress

Copper is an indispensable micronutrient that participates in a diverse range of physiological processes. However, plastocyanin, copper/zinc superoxide dismutase (CSD), and cytochrome c oxidase (COX) are highly prevalent Cu-rich proteins [91]. Cu-deficient conditions in plants are responsible for loss in agriculture production. Various approaches for modulating Cu homeostasis within plant have emerged, which involve the upregulation of Cu-miRNAs, eventually target mRNAs, and induce the release of Cu from protein to fulfill other essential requirements in plant [92]. miR398, miR397, miR408, and miR857 are the most significant miRNAs that respond in Cu deficiency [93]. In this process, Cu-responsive transcription factor SPL7 interacts with GTAC motifs in the *MIR398* promoter region; eventually turns on the degradation of *Cu/Zn-SOD*, *CCS1*, and *COX5b-1* mRNAs; and facilitates the transfer of Cu from CSDs to PC; Cu/Zn-SODs biological process switches to Fe-SOD (FSD1) [94]. Moreover, in the Cu-deficient conditions, the miR398-dependent downregulation of CCS1 facilitates Cu release from CSDs [95]. Several miRNA have been studied that control transcription factor in Cu-deficient conditions such as miR156 regulate SPL 3 (SQUAMOSA-PROMOTER BINDING PROTEIN-LIKE) transcription family member at posttranscriptional stage, activate in Cu-scarce condition, and act accordingly based on cues. RT-qPCR study for miR156, miR157 was conducted to study the expression level for different Cu concentration in media, and it was confirmed that miR156 and miR157 expressed more in shoot compared with in root [96].

4. Role of miRNAs in metal toxicity

Crop productivity is significantly hindered by metal toxicity, which encompasses both indispensable plant metals like copper, iron, zinc, and manganese, and nonessential metals such as cadmium (Cd), aluminum (Al), cobalt (Co), and mercury (Hg). An overarching consequence of elevated metal concentrations—be it aluminum, copper, cadmium, or mercury—is the suppression of root growth. The harmful impact of metal toxicity prompts the accumulation of reactive oxygen species, leading to detrimental effects on lipids, proteins, and DNA. The plant's reaction to metal toxicity involves intricate biological mechanisms that demand precise regulation at both transcriptional and posttranscriptional tiers [97].

Examinations of small RNA expression patterns in plants exposed to metal toxicities have illuminated distinct miRNA expression patterns and their corresponding target variations, hinting at potential regulatory or signaling functions. A significant proportion of predicted targets for metal-responsive, evolutionarily conserved miRNAs are transcription factors, primarily associated with plant development. When a particular miRNA experiences upregulation leading to target degradation, it suggests that the targeted element might act as a negative modulator in the plant's response to metal toxicity. Recent advancements in high-throughput genomic technologies and other genetic-genomic approaches have substantially expanded our understanding of miRNAs and their interactions with targets within signaling pathways concerning plant responses to metal toxicities. This growing body of knowledge has been gleaned from various plant species, further enhancing our insights into the intricate roles played by miRNAs in orchestrating responses to metal-induced stresses [98].

When it comes to miRNAs that respond to cadmium (Cd) toxicity, a set of conserved and new miRNAs have been found in rice seedlings that were exposed to Cd. Among the conserved miRNAs, miR160, miR164, and miR167, along with the novel Osa-miR602 and Osa-miR604, were discovered [99]. Notably, Osa-miR602 exhibited upregulation in rice roots after a 12-hour exposure to elevated Cd levels, with a xyloglucan endotransglucosylase/hydrolase as its predicted target. On the other hand, Osa-miR604, found to be upregulated in Cd-treated leaves for 6 hours, which led to the down regulation of a lipid transfer protein (LPT) [99]. In rice microarray data, the response to Cd stress revealed upregulation of miR528, while miR162, miR166, miR171, miR390, miR168, and miR156 families were downregulated [100]. Investigation into potential metal-responsive cis-acting elements highlighted the presence of a MRE-like sequence (5′-TGCGCNC-3′) in the promoter regions of most Cd-responsive miRNA genes [100].

In *B. napus* roots, Cd exposure for 8 hours led to downregulation of miR393, miR171, miR156, and miR396 [101]. Similarly, in the leaves of the model legume *M. truncatula*, Cd, Hg, and Al exposure triggered the upregulation of miR393, miR171, miR319, and miR529, while miR166 and miR398 were downregulated [102]. High-throughput small RNA sequencing unraveled the downregulation of miR159, miR160, miR319, miR396, and miR390 in response to aluminum (Al) toxicity [103]. More recently, a similar approach has identified miRNAs responsive to mercury (Hg) toxicity, including the upregulation of miR167, miR172, miR169, miR164, and miR395 families, while miR396, miR390, and miR171 were downregulated in this legume. Moreover, novel Hg-responsive miRNAs like miR2681, targeting transcripts coding for TIR-NBS-LRR disease resistance proteins, were uncovered in *M. truncatula* [102]. These findings underscore the intricate network of miRNA responses to diverse metal toxicities across various plant species.

5. Limitations of ncRNAs in plant-nutrient homeostasis

Nutrient uptake and metabolism are highly complex processes involving multiple genes, pathways, and regulatory mechanisms. Noncoding RNAs are just one layer of this complexity, and they may not be the primary regulators of nutrient uptake in many cases. Only ncRNAs do not act in isolation but are part of a larger regulatory network that includes transcription factors and other regulatory elements. Understanding how ncRNAs interact with these other components is crucial for a comprehensive understanding of nutrient uptake regulation. Our understanding of the role in specific ncRNAs in nutrient uptake is still evolving. The role of ncRNAs in regulation of nutrient transportation has not undergone comprehensive studies till date [104]. While some ncRNAs have been identified as regulators of nutrient-related genes, there is much we do not know about the intricacies of these regulatory networks.

miRNAs often target multiple genes, and a single miRNA can regulate the expression of several mRNAs. This lack of specificity can make it challenging to precisely control the expression of genes related to nutrient uptake without affecting other important cellular processes. Just for example, downregulation of miR169 in *Arabidopsis thaliana* affects the expression of multiple genes like nuclear transcription factor Y subunit-alpha (*NFYA*) genes, namely, *NFYA8*, *NFYA5*, and *NFYA2* [37]. Similarly, over-expression of miR169a strongly correlates with the inhibition of nitrate transport system (*NRT*) genes, *AtNRT2.1* and *AtNRT1.1*. So, this nonspecific binding of miRNA with multiple genes may create a problem for regulation of miRNA in nutrient uptake as well as transgenic development [37]. Secondly, redundancy creates a major problem for researchers for proper target identification. Many miRNAs have redundant functions, meaning that multiple miRNAs can regulate the same target genes. For example, in *M. truncatula*, multiple miRNAs, namely, miR2118, miR2109, and miR1507, act on NB-LRR genes for symbiotic interactions. This redundancy can compensate for the loss of one miRNA, making it difficult to observe significant phenotypic changes solely by manipulating a single miRNA [64]. In some studies, it was also observed that miRNA reduces the growth of lateral roots. miR167 serves to impede the accumulation of ARF8 (Auxin Response Factor 8) within the pericycle cell layer, which leads to reduction in the formation of lateral roots as well as N metabolic enzymes that are produced by the downstream of nitrification and absorption [105, 106]. Certain noncoding RNAs exhibit pleiotropic morphology effects on traits in transgenic situations, raising doubts about the precision of miRNA transformation for a specific event [107]. Conducting experiments to manipulate miRNAs and study their effects on nutrient uptake can be technically challenging. When miRNAs are subjected to cytoplasm and released from endosome, off-target effects play a major concern. These noncoding RNAs are produced to target various nutrient uptake pathways by imperfect hybridization with 3′ UTRs; thus, they might cause undesirable gene silencing of other genes. This off-target gene silencing can result in potential harmful effects and diminished therapeutic efficacy. The fact that a single miRNA could potentially affect multiple mRNA targets necessitates careful consideration, as it suggests the potential for unforeseen side effects [108].

6. Conclusion

Exploration into the realm of microRNAs (miRNAs) has illuminated their pivotal role in shaping the landscape of nutrient uptake within plants. The presented studies intricately elucidate the symbiotic interplay between miRNAs and diverse nutrient stresses, encompassing both micro- and macronutrient deficiencies. These revelations underscore the profound influence of miRNAs in orchestrating gene expression, thereby exerting a discernible impact on critical facets such as ion transport, metabolic adaptability, and acclimatization to nutrient-scarce conditions. The holistic comprehension of miRNA-mediated mechanisms beckons us toward promising avenues for augmenting crop productivity and bolstering resilience amidst the challenges posed by nutrient-related constraints. A deeper plunge into the intricate tapestry of miRNA-nutrient stress interactions harbors profound potential for devising innovative strategies that fine-tune nutrient uptake and utilization in plants. This endeavor aligns seamlessly with the objectives of sustainable agriculture and the global quest for food security. Furthermore, it is imperative to shift our focus toward unraveling the nuanced involvement of miRNAs in nutritional stress responses. This endeavor necessitates a keen observation of the phenotypic and physiological shifts induced by these regulatory processes. While the spotlight often falls on miRNAs, it is worth acknowledging that nutritional stress can also extend its reach to transposon-derived RNAs and tRNA-derived RNAs, with their precise physiological roles yet to be unraveled. In the pursuit of fortifying future food security, harnessing miRNA-centric strategies emerges as a vital linchpin. Based on the potential of miRNA-based interventions, it might be possible to make crop varieties that are more productive and resistant to both natural and man-made stresses.

Acknowledgements

We are sincerely grateful to the esteemed Vice Chancellor of G.B. Pant University of Agriculture and Technology (GBPUAT), Dean CBSH and Hrad MBGE, for their unwavering support and dedication in making this endeavor possible.

Funding

This research did not receive any specific grant from funding agencies in the public, commercial, or not-for-profit sectors.

Conflict of interest

The authors declare no conflict of interest.

Author details

Jajati Keshari Nayak*, Rashmi Chauhan, Sundip Kumar and Sharat Prabhakaran
Department of Molecular Biology and Genetic Engineering, G.B. Pant University of Agriculture and Technology, Pantnagar, Uttrakhand, India

*Address all correspondence to: jajatikesharinayak00@gmail.com

References

[1] Marschner H, Mineral Nutrition of Higher Plants. Cambridge: Academic Press; 1995

[2] Schachtman DP, Reid RJ, Ayling SM. Phosphorus uptake by plants: From soil to cell. Plant Physiology. 1998;**116**(2):447-453

[3] Sperotto RA, Ricachenevsky FK, Williams LE, Vasconcelos MW, Menguer PK. From soil to seed: Micronutrient movement into and within the plant. Frontiers in Plant Science. 2014;**5**:438

[4] Vigani G, Zocchi G, Bashir K, Philippar K, Briat JF. Signals from chloroplasts and mitochondria for iron homeostasis regulation. Trends in Plant Science. 2013;**18**(6):305-311

[5] Jung JK, McCouch S. Getting to the roots of it: Genetic and hormonal control of root architecture. Frontiers in Plant Science. 2013;**4**:186

[6] Jones Rhoades MW, Bartel DP. Computational identification of plant microRNAs and their targets, including a stress-induced miRNA. Molecular Cell. 2004;**14**:787-799

[7] Schauer SE, Jacobsen SE, Meinke DW, Ray A. DICER-LIKE1: Blind men and elephants in Arabidopsis development. Trends in Plant Science. 2002;**7**(11):487-491

[8] Contreras-Cubas C, Palomar M, Arteaga-Vázquez M, Reyes JL, Covarrubias AA. Non-coding RNAs in the plant response to abiotic stress. Planta. 2012;**236**:943-958

[9] Nozawa M, Miura S, Nei M. Origins and evolution of microRNA genes in drosophila species. Genome Biology and Evolution. 2010;**2**:180-189

[10] Zhu L, Liu X, Pu W, Peng Y. tRNA-derived small non-coding RNAs in human disease. Cancer Letters. 2018a;**419**:1-7

[11] Megel C, Hummel G, Lalande S, Ubrig E, Cognat V, Morelle G, et al. Plant RNases T2, but not Dicer-like proteins, are major players of tRNA-derived fragments biogenesis. Nucleic Acids Research. 2019;**47**(2):941-952

[12] Waititu JK, Zhang C, Liu J, Wang H. Plant non-coding RNAs: Origin, biogenesis, mode of action and their roles in abiotic stress. International Journal of Molecular Sciences. 2020;**21**:8401

[13] Bonnet E, Van de Peer Y, Rouzé P. The small RNA world of plants. New Phytologist. 2006;**171**(3):451-468

[14] Yang X, Li L. Analyzing the microRNA transcriptome in plants using deep sequencing data. Biology. 2012;**1**(2):297-310

[15] Liu X, Hao L, Li D, Zhu L, Hu S. Long non-coding RNAs and their biological roles in plants. Genomics, Proteomics & Bioinformatics. 2015;**13**(3):137-147

[16] Yu B, Yang Z, Li J, Minakhina S, Yang M, Padgett RW, et al. Methylation as a crucial step in plant microRNA biogenesis. Science. 2005;**307**(5711):932-935

[17] Bologna NG, Iselin R, Abriata LA, Sarazin A, Pumplin N, Jay F, et al. Nucleo-cytosolic shuttling of ARGONAUTE1 prompts a revised model of the plant microRNA pathway. Molecular Cell. 2018;**69**(4):709-719

[18] Millar AA. The function of miRNAs in plants. Plants. 2020;**9**(2):198

[19] Sablok G, Zhao H, Sun X. Plant circular RNAs (circRNAs): Transcriptional regulation beyond miRNAs in plants. Molecular Plant. 2016;**9**(2):192-194

[20] Alves CS, Vicentini R, Duarte GT, Pinoti VF, Vincentz M, Nogueira FT. Genome-wide identification and characterization of tRNA-derived RNA fragments in land plants. Plant Molecular Biology. 2017;**93**:35-48

[21] Cognat V, Morelle G, Megel C, Lalande S, Molinier J, Vincent T, et al. The nuclear and organellar tRNA-derived RNA fragment population in Arabidopsis thaliana is highly dynamic. Nucleic Acids Research. 2017;**45**(6):3460-3472

[22] Axtell MJ, Meyers BC. Revisiting criteria for plant microRNA annotation in the era of big data. The Plant Cell. 2018;**30**(2):272-284

[23] Rogers K, Chen X. Biogenesis, turnover, and mode of action of plant microRNAs. The Plant Cell. 2013;**25**(7):2383-2399

[24] Yu Y, Jia T, Chen X. The 'how' and 'where' of plant micro RNAs. New Phytologist. 2017;**216**(4):1002-1017

[25] Kong X, Yang M, Le BH, He W, Hou Y. The master role of siRNAs in plant immunity. Molecular Plant Pathology. 2022;**23**(10):1565-1574

[26] Chen M, Penfield S. Feedback regulation of COOLAIR expression controls seed dormancy and flowering time. Science. 2018;**360**:1014-1017

[27] Song Y, Xuan A, Bu C, Ci D, Tian M, Zhang D. Osmotic stress-responsive promoter upstream transcripts (PROMPTs) act as carriers of MYB transcription factors to induce the expression of target genes in Populus simonii. Plant Biotechnology Journal. 2019;**17**:164-177

[28] Wang HLV, Chekanova JA. Long noncoding RNAs in plants. Long non coding. RNA Biology. 2017;**1008**:133-154

[29] Matsumoto A, Nakayama KI. Hidden peptides encoded by putative noncoding RNAs. Cell Structure and Function. 2018;**43**:75-83

[30] Hawkesford M, Horst W, Kichey T, Lambers H, Schjoerring J, Møller IS, et al. Chapter 6 – Functions of macronutrients. In: Marschner P, editor. Marschner's Mineral Nutrition of Higher Plants. 3rd ed. San Diego, CA: Academic Press; 2012. pp. 135-189. DOI: 10.1016/B978-0-12-384905-2.00006-6

[31] Crawford NM, Forde BG. Molecular and developmental biology of inorganic nitrogen nutrition. Arabidopsis Book. 2002;**1**:e0011

[32] Wang Y, Zhang C, Hao Q, Sha A, Zhou R, Zhou X, et al. Elucidation of miRNAs-mediated responses to low nitrogen stress by deep sequencing of two soybean genotypes. PLoS One. 2013;**8**:e67423

[33] Xu Z, Zhong S, Li X, Li W, Rothstein SJ, Zhang S, et al. Genome-wide identification of microRNAs in response to low nitrate availability in maize leaves and roots. PLoS One. 2011;**6**:e28009

[34] Trevisan S, Nonis A, Begheldo M, Manoli A, Palme K, Caporale G, et al. Expression and tissue-specific localization of nitrate-responsive miRNAs in roots of maize seedlings. Plant & Cell Environment. 2012;**35**:1137-1155

[35] Zhao M, Ding H, Zhu JK, Zhang F, Li WX. Involvement of miR169 in the nitrogen-starvation responses in Arabidopsis. New Phytologist. 2011;**190**(4):906-915

[36] Pant BD, Musialak-Lange M, Nuc P, May P, Buhtz A, Kehr J, et al.

Identification of nutrient-responsive Arabidopsis and rapeseed microRNAs by comprehensive real-time polymerase chain reaction profiling and small RNA sequencing. Plant Physiology. 2009;**150**(3):1541-1555

[37] Liang G, He H, Yu D. Identification of nitrogen starvation-responsive microRNAs in *Arabidopsis thaliana*. PLoS One. 2012;7:e48951

[38] Gifford ML, Dean A, Gutierrez RA, Coruzzi GM, Birnbaum KD. Cell-specific nitrogen responses mediate developmental plasticity. National Academy of Sciences of the United States of America. 2008;**105**:803-808

[39] Ordóñez-Baquera PL, González-Rodríguez E, Aguado-Santacruz GA, Rascón-Cruz Q, Conesa A, Moreno-Brito V, et al. Identification of miRNA from Bouteloua gracilis, a drought tolerant grass, by deep sequencing and their in silico analysis. Computational Biology and Chemistry. 2017;**66**:26-35

[40] Vidal EA, Moyano TC, Riveras E, Contreras-López O, Gutiérrez RA. Systems approaches map regulatory networks downstream of the auxin receptor AFB3 in the nitrate response of Arabidopsis thaliana roots. National Academy of Sciences of the United States of America. 2013;**110**(31):12840-12845

[41] Yan Y, Wang H, Hamera S, Chen X, Fang R. miR444a has multiple functions in the rice nitrate-signaling pathway. The Plant Journal. 2014;**78**(1):44-55

[42] Zeng J, Ye Z, He X, Zhang G. Identification of microRNAs and their targets responding to low-potassium stress in two barley genotypes differing in low-K tolerance. Journal of Plant Physiology. 2019;**234**:44-53

[43] Rey-Burusco MF, Daleo GR, Feldman ML. Identification of potassium phosphite responsive miRNAs and their targets in potato. PLoS One. 2019;**14**(9):e0222346

[44] Zeng H, Zhang X, Ding M, Zhu Y. Integrated analyses of miRNAome and transcriptome reveal zinc deficiency responses in rice seedlings. BMC Plant Biology. 2019;**19**(1):1-18

[45] Zhao Y, Xu K, Liu G, Li S, Zhao S, Liu X, et al. Global identification and characterization of miRNA family members responsive to potassium deprivation in wheat (*Triticum aestivum* L.). Scientific Reports. 2020;**10**(1):15812

[46] Kirkby EA. Maximizing calcium uptake by plants. Communications in Soil Science and Plant Analysis. 1979;**10**(1-2):89-113

[47] Guo HS, Xie Q, Fei JF, Chua NH. MicroRNA directs mRNA cleavage of the transcription factor NAC1 to downregulate auxin signals for Arabidopsis lateral root development. The Plant Cell. 2005;**17**:1376-1386. DOI: 10.1105/tpc.105.030841

[48] Andrews M, Andrews ME. Specificity in legume-rhizobia symbioses. International Journal of Molecular Sciences. 2017;**18**(4):705

[49] Clúa J, Roda C, Zanetti ME, Blanco FA. Compatibility between legumes and rhizobia for the establishment of a successful nitrogen-fixing symbiosis. Genes. 2018;**9**(3):125

[50] Oldroyd GE. Speak, friend, and enter: Signalling systems that promote beneficial symbiotic associations in plants. Nature Reviews Microbiology. 2013;**11**(4):252-263

[51] Ferguson BJ, Indrasumunar A, Hayashi S, Lin MH, Lin YH, Reid DE, et al. Molecular analysis of legume nodule development and autoregulation.

Journal of Integrative Plant Biology. 2010;**52**(1):61-76

[52] Peláez-Vico MA, Bernabéu-Roda L, Kohlen W, Soto MJ, López-Ráez JA. Strigolactones in the rhizobium-legume symbiosis: Stimulatory effect on bacterial surface motility and down-regulation of their levels in nodulated plants. Plant Science. 2016;**245**:119-127

[53] Lelandais-Brière C, Naya L, Sallet E, Calenge F, Frugier F, Hartmann C, et al. Genome-wide Medicago truncatula small RNA analysis revealed novel microRNAs and isoforms differentially regulated in roots and nodules. The Plant Cell. 2009;**21**(9):2780-2796

[54] Joshi T, Yan Z, Libault M, Jeong DH, Park S, Green PJ, et al. Prediction of novel miRNAs and associated target genes in Glycine max. BMC Bioinformatics. 2010;**11**:1-9

[55] Sós-Hegedűs A, Domonkos Á, Tóth T, Gyula P, Kaló P, Szittya G. Suppression of NB-LRR genes by miRNAs promotes nitrogen-fixing nodule development in *Medicago truncatula*. Plant & Cell Environment. 2020;**43**:1117-1129

[56] Liang WW, Huang JH, Li CP, Yang LT, Ye X, Lin D, et al. MicroRNA-mediated responses to long-term magnesium-deficiency in *Citrus sinensis* roots revealed by Illumina sequencing. BMC Genomics. 2017;**18**(1):1-16

[57] Ma CL, Qi YP, Liang WW, Yang LT, Lu YB, Guo P, et al. MicroRNA regulatory mechanisms on *Citrus sinensis* leaves to magnesium-deficiency. Frontiers in Plant Science. 2016;**7**:201

[58] Astolfi S, Celletti S, Vigani G, Mimmo T, Cesco S. Interaction between Sulfur and iron in plants. Frontiers in Plant Science. 2021;**12**:670308

[59] Dietzen C, Koprivova A, Whitcomb SJ, Langen G, Jobe TO, Hoefgen R, et al. The transcription factor EIL1 participates in the regulation of sulfur-deficiency response. Plant Physiology. 2020;**184**(4): 2120-2136

[60] Rotte C, Leustek T. Differential subcellular localization and expression of ATP sulfurylase and 5′-adenylylsulfate reductase during ontogenesis of Arabidopsis leaves indicates that cytosolic and plastid forms of ATP sulfurylase may have specialized functions. Plant Physiology. 2000;**124**:715-724

[61] Liang G, Yang F, Yu D. MicroRNA395 mediates regulation of sulfate accumulation and allocation in Arabidopsis thaliana. The Plant Journal. 2010;**62**:1046-1057

[62] Çelik Ö, Akdaş EY. Tissue-specific transcriptional regulation of seven heavy metal stress-responsive miRNAs and their putative targets in nickel indicator castor bean (*R. Communis* L.) plants. Ecotoxicology and Environmental Safety. 2019;**170**:682-690

[63] Huang SQ, Xiang AL, Che LL, Chen S, Li H, Song JB, et al. A set of miRNAs from Brassica napus in response to sulphate deficiency and cadmium stress. Plant Biotechnology Journal. 2010;**8**:887-899

[64] Kong WW, Yang ZM. Identification of iron-deficiency responsive microRNA genes and cis-elements in Arabidopsis. Plant Physiology and Biochemistry. 2010;**48**(2-3):153-159

[65] Zhao H, Sun R, Albrecht U, Padmanabhan C, Wang A, Coffey MD, et al. Small RNA profiling reveals phosphorus deficiency as a contributing factor in symptom expression for citrus huanglongbing disease. Molecular Plant. 2013;**6**(2):301-310

[66] Huang W, Liu J, Wang YP, Zhou G, Han T, Li Y. Increasing phosphorus limitation along three successional forests in southern China. Plant and Soil. 2013;**364**:181-191

[67] Branscheid A, Devers EA, May P, Krajinski F. Distribution pattern of small RNA and degradome reads provides information on miRNA gene structure and regulation. Plant Signaling and Behavior. 2011;**6**:1609-1611

[68] Devers EA, Branscheid A, May P, Krajinski F. Stars and symbiosis: microRNA-and microRNA*-mediated transcript cleavage involved in arbuscular mycorrhizal symbiosis. Plant Physiolgy. 2011;**156**:1990-2010

[69] Pettigrew WT. Potassium influences on yield and quality production for maize, wheat, soybean and cotton. Physiologia Plantarum. 2008;**133**(4):670-681

[70] Wang Y, Wu WH. Genetic approaches for improvement of the crop potassium acquisition and utilization efficiency. Current Opinion in Plant Biology. 2015;**25**:46-52

[71] Hsieh LC, Lin SI, Shih ACC, Chen JW, Lin WY, Tseng CY, et al. Uncovering small RNA-mediated responses to phosphate deficiency in Arabidopsis by deep sequencing. Plant Physiology. 2009;**151**(4):2120-2132

[72] Numan M, Guo W, Choi SC, Wang X, Du B, Jin W, et al. Analysis of miRNAs responsive to long-term calcium deficiency in tef (*Eragrostis tef* (Zucc.) trotter). Plant Direct. 2022;**6**(5):e400

[73] Chen H, Yang Q, Chen K, Zhao S, Zhang C, Pan R, et al. Integrated microRNA and transcriptome profiling reveals a miRNA-mediated regulatory network of embryo abortion under calcium deficiency in peanut (*Arachis hypogaea L.*). BMC Genomics. 2019;**20**:1-17

[74] Paul S, Gayen D, Datta SK, Datta K. Analysis of high iron rice lines reveals new miRNAs that target iron transporters in roots. Journal of Experimental Botany. 2016;**67**(19):5811-5824

[75] Islam W, Tauqeer A, Waheed A, Zeng F. MicroRNA mediated plant responses to nutrient stress. International Journal of Molecular Sciences. 2022;**23**(5):2562

[76] Broadley MR, White PJ, Hammond JP, Zelko I, Lux A. Zinc in plants. New Phytologist. 2007;**173**(4):677-702

[77] Li Y, Zhang Y, Shi D, Liu X, Qin J, Ge Q, et al. Spatial-temporal analysis of zinc homeostasis reveals the response mechanisms to acute zinc deficiency in Sorghum bicolor. New Phytologist. 2013;**200**(4):1102-1115

[78] Shi DQ, Zhang Y, Li YL, Jin XU. Identification of zinc deficiency-responsive microRNAs in Brassica juncea roots by small RNA sequencing. Journal of Integrative Agriculture. 2013;**12**(11):2036-2044

[79] Zhao J, Wang W, Zhou H, Wang R, Zhang P, Wang H, et al. Manganese toxicity inhibited root growth by disrupting auxin biosynthesis and transport in Arabidopsis. Frontiers in Plant Science. 2017;**8**:272

[80] Alejandro S, Höller S, Meier B, Peiter E. Manganese in plants: From acquisition to subcellular allocation. Frontiers in Plant Science. 2020;**11**:300

[81] Sabelli PA, Larkins BA. The contribution of cell cycle regulation to endosperm development. Sexual Plant Reproduction. 2009a;**22**:207-219

[82] Yang SX, Deng H, Li MS. Manganese uptake and accumulation

in a woody hyperaccumulator, Schima superba. Plant Soil Environment. 2008;**54**(10):441-446

[83] Valdés-López O, Yang SS, Aparicio-Fabre R, Graham PH, Reyes JL, Vance CP, et al. MicroRNA expression profile in common bean (Phaseolus vulgaris) under nutrient deficiency stresses and manganese toxicity. New Phytologist. 2010;**187**(3):805-818

[84] Gong J, Li D, Li H, Zhou H, Xu J. Identification of manganese-responsive microRNAs in Arabidopsis by small RNA sequencing. Czech Journal of Genetics and Plant Breeding. 2019;**55**(2):76-82

[85] Shireen F, Nawaz MA, Chen C, Zhang Q, Zheng Z, Sohail H, et al. Boron: Functions and approaches to enhance its availability in plants for sustainable agriculture. International Journal of Molecular Sciences. 2018;**19**(7):1856

[86] Ozhuner E, Eldem V, Ipek A, Okay S, Sakcali S, Zhang B, et al. Boron stress responsive microRNAs and their targets in barley. PLoS One. 2013;**8**(3):e59543

[87] Jyothi MN, Usha S, Suchithra B, Ulfath TS, Devaraj VR, Babu RN. Boron toxicity induces altered expression of miRNAs in French bean (*Phaseolus vulgaris L.*). Journal of Applied Biology and Biotechnology. 2018;**6**(6):1-10

[88] Lu YB, Yang LT, Qi YP, Li Y, Li Z, Chen YB, et al. Identification of boron-deficiency-responsive microRNAs in Citrus sinensis roots by Illumina sequencing. BMC Plant Biology. 2014;**14**(1):1-16

[89] Jin LF, Liu YZ, Yin XX, Peng SA. Transcript analysis of citrus miRNA397 and its target LAC7 reveals a possible role in response to boron toxicity. Acta Physiologiae Plantarum. 2016;**38**:1-7

[90] Kayihan DS, Kayihan C, Çiftçi YÖ. Moderate level of toxic boron causes differential regulation of microRNAs related to jasmonate and ethylene metabolisms in Arabidopsis thaliana. Turkish Journal of Botany. 2019;**43**(2):167-172

[91] Beauclair L, Yu A, Bouché N. microRNA-directed cleavage and translational repression of the copper chaperone for superoxide dismutase mRNA in Arabidopsis. The Plant Journal. 2010;**62**(3):454-462

[92] Shabbir Z, Sardar A, Shabbir A, Abbas G, Shamshad S, Khalid S, et al. Copper uptake, essentiality, toxicity, detoxification and risk assessment in soil-plant environment. Chemosphere. 2020;**259**:127436

[93] Perea-García A, Andrés-Bordería A, Huijser P, Peñarrubia L. The copper-microrna pathway is integrated with developmental and environmental stress responses in *arabidopsis thaliana*. International Journal of Molecular Sciences. 2021;**22**:9547

[94] Yamasaki H, Hayashi M, Fukazawa M, Kobayashi Y, Shikanai T. SQUAMOSA promoter binding protein–like7 is a central regulator for copper homeostasis in Arabidopsis. The Plant Cell. 2009;**21**(1):347-361

[95] Li J, Song Q, Zuo ZF, Liu L. MicroRNA398: A master regulator of plant development and stress responses. International Journal of Molecular Sciences. 2022;**23**(18):10803

[96] Wang JW, Czech B, Weigel D. miR156-regulated SPL transcription factors define an endogenous flowering pathway in arabidopsis thaliana. Cell. **138**(4):738-749

[97] Mendoza-Soto AB, Sánchez F, Hernández G. MicroRNAs as regulators

in plant metal toxicity response. Frontiers in Plant Science. 2012;**3**:105

[98] Zhou ZS, Zeng HQ, Liu ZP, Yang ZM. Genome-wide identification of *Medicago truncatula* microRNAs and their targets reveals their differential regulation by heavy metal. Plant, Cell & Environment. 2012;**35**(1):86-99

[99] Huang SQ, Peng J, Qiu CX, Yang ZM. Heavy metal-regulated new microRNAs from rice. Journal of Inorganic Biochemistry. 2009;**103**:282-287

[100] Ding Y, Shi Y, Yang S. Advances and challenges in uncovering cold tolerance regulatory mechanisms in plants. New Phytologist. 2019;**222**(4):1690-1704

[101] Xie FL, Huang SQ, Guo K, Xiang AL, Zhu YY, Nie L, et al. Computational identification of novel microRNAs and targets in *Brassica napus*. FEBS Letters. 2007;**581**:1464-1474

[102] Zhou ZS, Huang SJ, Yang ZM. Bioinformatic identification and expression analysis of new microRNAs from *Medicago truncatula*. Biochemical and Biophysical Research Communications. 2008;**374**:538-542

[103] Chen RF, Zhang FL, Zhang QM, Sun QB, Dong XY, Shen RF. Aluminium-phosphorus interactions in plants growing on acid soils: Does phosphorus always alleviate aluminium toxicity? Journal of the Science of Food and Agriculture. 2012;**92**(5):995-1000

[104] Paul S, Datta SK, Datta K. miRNA regulation of nutrient homeostasis in plants. Frontiers in Plant Science. 2015;**6**:232

[105] Yousuf PY, Shabir PA, Hakeem KR. miRNAomic approach to plant nitrogen starvation. International Journal of Genomics. 2021;**2021**:8560323

[106] Arora S, Singh AK, Chaudhary B. Target-mimicry based miRNA167-diminution ameliorates cotton somatic embryogenesis via transcriptional biases of auxin signaling associated miRNAs and genes. Plant Cell, Tissue and Organ Culture. 2020;**141**:511-531

[107] Ferdous J, Sanchez-Ferrero JC, Langridge P, Milne L, Chowdhury J, Brien C, et al. Differential expression of microRNAs and potential targets under drought stress in barley. Plant, Cell & Environment. 2017;**40**(1):11-24

[108] Suter SR, Ball-Jones A, Mumbleau MM, Valenzuela R, Ibarra-Soza J, Owens H, et al. Controlling miRNA-like off-target effects of an siRNA with nucleobase modifications. Organic & Biomolecular Chemistry. 2017;**15**(47):10029-10036